The Prince of Mathematics

The Prince of Mathematics
Carl Friedrich Gauss

M.B.W. Tent

A K Peters, Ltd.
Wellesley, Massachusetts

Editorial, Sales, and Customer Service Office
A K Peters, Ltd.
888 Worcester Street, Suite 230
Wellesley, MA 02482
www.akpeters.com

Copyright ©2006 by A K Peters, Ltd.

All rights reserved. No part of the material protected by this copyright notice may be reproduced or utilized in any form, electronic or mechanical, including photocopying, recording, or by any information storage and retrieval system, without written permission from the copyright owner.

Library of Congress Cataloging-in-Publication Data
Tent, M. B. W. (Margaret B. W.), 1944–
  The prince of mathematics : Carl Friedrich Gauss / M.B.W. Tent.
    p. cm.
  Includes index.
  ISBN 13: 978-1-56881-261-8 (alk. paper)
  ISBN 10: 1-56881-261-2 (alk. paper)
    1. Gauss, Carl Friedrich, 1777–1855. 2. Mathematicians—Germany—Biography. 3. Astronomers—Germany—Biography. 4. Cartographers—Germany—Biography. I. Title.
  QA29.G3T46   2005
  510/.92—dc22   2005051464

Cover images: Top left: portrait of Gauss by J.C.A. Schwartz, 1803. Bottom left: lithograph of Gauss on the terrace of the Observatory in Göttingen by Eduard Ritmüller (no date). The text in the top right comes from the marriage certificate of Gauss and Johanna Osthoff, 1805.

The illustrations on the cover and on pages 5, 52, 89, 94, 96, 131, 146, 174, and 188 have been provided courtesy of the Gauss Society Göttingen. The photograph on page 220 was taken by Dennis Martin Wittmann. The top on page 32 is owned by and being held by Tom Fox.

Printed in India
10 09 08 07 06   10 9 8 7 6 5 4 3 2 1

To my father
Raymond Wyman
my mentor for sixty years

Contents

Foreword — xi

Preface — xv

Acknowledgements — xvii

1 Child Prodigy (1777–1788) — 1

Counting — 3

Sums — 11

The Duke of Braunschweig — 21

The Schoolroom — 25

Arithmetic — 33

Early Mathematics — 41

II The Duke's Protégé (1788–1798) — 49

Royal Patronage — 51

Gymnasium: High School — 63

Independent Study of Mathematics — 75

The Regular 17-gon — 81

Mathematical Journal — 95

Number Theory — 101

III Gifted Astronomer, Father of a Young Family (1798–1814) — 113

Carl Friedrich Gauss, PhD — 115

The Planetoid Ceres — 125

A Wife and a Child — 133

The Duke and St. Petersburg — 139

Professor of Astronomy — 149

Tragedy — 153

Marriage to Minna Waldeck — 159

CONTENTS

The Trip to Munich	165
The New Observatory	173
Gauss' Mother	179

IV Surveyor of Hannover, Father of a Growing Family (1815–1832) — **185**

Surveying	187
Summers on the Road	195
Trouble with Eugen	203

V Magnetic Professor, Prince of Mathematics (1833–1855) — **211**

Non-Euclidean Geometry	213
Magnetism	217
The Göttingen Seven	225
Looking Back	233
Index	239

Foreword

"Nothing has been done if something remains to be done," Carl Friedrich Gauss, one of the greatest scientists of all times, used to say. Which of course means that a problem (at least a mathematical problem) has to be finally and comprehensively solved, else it is considered unsolved. Born the son of a workman, Gebhard Gauss, and his wife Dorothea (née Benze) in Brunswick in 1777, Carl Friedrich's talents became evident at an early age. In 1791 he was granted a sponsorship by the Duke of Brunswick, Carl Wilhelm Ferdinand. This enabled Gauss to buy monographs, to attend the high school at Brunswick, and to study mathematics and astronomy "abroad," namely at Göttingen University in the state of Hanover, from 1795–1798. Thereafter Gauss returned to Brunswick and got his doctoral degree in mathematics at the then Brunswick State University in Helmstedt. His *Disquisitiones Arithmeticae*, which was published in 1801, made Gauss famous, in particular among the elite of mathematicians. But an even bigger breakthrough came when Gauss was able to mathematically rediscover the first minor planet, Ceres, which had vanished from sight after a brief period of observation, in 1801. This made Gauss famous worldwide overnight.

In 1807 Gauss became professor of astronomy at Göttingen and director of Göttingen University Observatory. The latter had already gained international reputation through the work of Tobias Mayer (1723-1762), a genius whom Gauss constantly referred to as "*immortalis* Mayer." Thus, Gauss did not become a professional mathematician but—largely in accordance with his intentions—a professional astronomer. He spent the rest of his life, until his death in 1855, in that position at Göttingen and—like most astronomers of the time—was not only working but also living in his observatory. As was customary within the philosophical faculty, Gauss also worked and lectured in other fields, in particular mathematics, land surveying, geodesy, and physics, and made highly important contributions to each of these fields. Together with his much younger colleague, Wilhelm Weber (1804-1891), Gauss invented and operated the first electromagnetic telegraph of the world. He measured and computed the earth's magnetic field in great detail and developed new mathematical methods (including, for instance, what is now known as the Fast Fourier Transform, the Universal Transverse Mercator coordinates, and so on). Also, he taught a fairly large number of astronomers who later became famous, among them Heinrich Christian Schumacher (1780–1850), founder of the *Astronomische Nachrichten*, and Benjamin Apthorp Gould (1824–1896), founder of the *Astronomical Journal* (both journals still exist).

≡

One bright sunny day in June 2004 I was lucky to meet Margaret Tent at Göttingen, who, together with her husband, was on an extended research trip "on the heels of Gauss." She told me about her pupils at Altamont School in Birmingham, Alabama, and her plans to write a novel about Gauss that might be comprehensible also for children, conveying the joy of mathematics to them. I was electrified, because a few weeks earlier, during the preparations for the "Gauss Year" 2005 at Göttingen, we had been looking desperately for a novel about Gauss, but there was none. Oddly enough, as became particularly evident during the "Einstein Year" 2005, there are hundreds of biographies and monographs about Albert Einstein, but not more than a few about Gauss, most of them outdated. The most important among those, published in English, is G. Waldo Dunnington's *Gauss: Titan of Science*, which first appeared in New York in 1955 and was reprinted (augmented but otherwise unchanged) in Washington in 2004. This may have to do with the fact that mathematics and physics are difficult to understand: Almost no one really understands Einstein's theories, but—in contrast to Gauss—Einstein was living at a time where radio interviews, news reels, etc. conveyed his opinions. Also, Einstein publicly mingled in politics, which Gauss tried to avoid. Like Einstein, Gauss was a very honest character who almost never put his own interest in the forefront and who almost always lent support to friends and colleagues in trouble. But he kept loyal to his government (which had fostered him and his career) even when his personal fate was affected, as in the case of the "Göttingen Seven," who had protested against the King and lost their jobs (among them Weber and Ewald, Gauss' son-in-law).

When Margaret sent me a draft of her book, I was again electrified; I had never seen such a sympathetic understanding and detailed description of what presumably went on in Gauss's head:

fascination with nature, simple but straightforward explanations of mathematical relations and theorems, insights into the behavior of numbers, the motion of planets, and so on—that is, the joy and fun of mathematics, its application to real things, and one's ability to understand it. As Gauss once said, nature and its phenomena must be describable by mathematics, otherwise we won't be able to understand them. Though this book is a novel and does not pretend to be anything else, it—at least fictitiously—comes very close to a true autobiography, which Gauss himself never had an opportunity to write.

Axel Wittmann
Gauss Society Göttingen

Preface

One morning in 1992, shortly after I started teaching at the Altamont School in Birmingham, Alabama, I found a message scrawled on my chalkboard. "Mathematics is the queen of science, and number theory is the queen of mathematics." David Goldenberg and Bobo Blankson, two seventh grade boys, explained that they had found it in a book in our school library: Eric Temple Bell's *Men of Mathematics*. When I went to the library to look, I discovered a wonderful narrative on the history of mathematics. I was hooked.

In many ways this biography is an outgrowth of my interaction with students. They wanted to know more, so I needed to learn more. Since it was Carl Friedrich Gauss who made the statement about number theory, it is only natural that Gauss became the subject of this biographical account.

This narrative of Gauss' life is based on the stories Gauss told about himself and letters and descriptions that have come down to us. The vignettes and conversations are based as closely as possible on reports of what actually happened. The stories of three-year-old Gauss correcting his father's arithmetic and later falling into the canal and of ten-year-old Gauss figuring the sum of the first 100 counting numbers in school are all classics that have been told about Gauss many times over the years.

I hope that this story of the most mathematical child who ever lived and who then grew up to be recognized throughout Europe as the Prince of Mathematics will inspire readers to explore the world of mathematics. If it accomplishes that, I will be happy.

Acknowledgements

Many people were involved in the construction of this biographical account; colleagues, friends, and relatives all helped me in countless ways. I am grateful to all of them.

The Altamont School's grant from the Edward E. Ford Foundation provided the funding for my first trip to Göttingen as I began this project. At the Altamont School, Mary Martin, Shilpa Reddy, Danielle Wattleton, Daniel Hollingshead, and Clare Gamlin read the manuscript from beginning to end and provided me with important feedback. Katherine Berdy and Mary Jim Quillen read sections to help me as I framed the story. Devin Reeves, Trudy Loop, Eddy Dunn, and Mary Gray Hunter provided valuable technical support when I needed it. Jonathan Kentros, my student worker, gave me technical, stylistic, and mathematical help many times over the last two years.

The atmosphere of learning at Altamont has also played a role in my writing. Our students are curious, my colleagues are supportive, and everyone reads a great deal. Not everyone has the privilege of living and working in such intellectual surroundings. Our headmasters, both past and present, helped me as I struggled to meet deadlines at the same time that I also taught full time.

Many friends also provided important help. Catalina Herrera read the manuscript carefully and provided insightful comments. The Koch family in Berlin has provided wonderful support to our entire family over the years, and Christian Koch helped me locate Papen's Atlas in the city library of Berlin.

Roy Roddam loaned me an antique medical textbook with information on treating tuberculosis in the nineteenth century, and Donna Kentros helped me understand gestational diabetes and its effects on a mother and child. Markus Kupka procured a variety of resources for me. Friedel Rehkop, the city archivist in Dransfeld, provided many resources as well.

Axel Wittman, director of the observatory in Gttingen and Geschäftsführer of the Gauss-Gesellschaft, had a pivotal role in this work, showing me around the city of Göttingen, opening the facilities at the observatory to me, providing many photographs for the book, and reading the entire manuscript and checking for historical accuracy.

My husband, Jim, and my children, John and Virginia, have also been consistently supportive in many ways as this project moved forward. Jim and John helped with the historical framework for Gauss' life, filling in some important details of life in Germany at the time and helping me avoid anachronisms in the narrative. Virginia read and edited multiple drafts of the manuscript with intelligence and care. Above all, they supported and encouraged me in this huge project.

I am grateful to all these people and any others that I may have neglected to mention. You all made it possible for me to complete this project. Thank you.

Part I

Child Prodigy
1777–1788

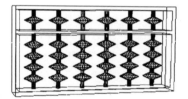

Part I

Child Prodigy
1777–1788

Counting

Carl Friedrich Gauss' father, a stonemason, was paying his workmen at the end of the day: "Let's see, *Herr* [Mr.] Braun, that's 34 pennies plus 29 pennies plus 19 pennies and ... um ... that makes a total of 76 pennies."

Young Carl, watching and listening from the front step, said clearly, "No, Father, that isn't right. It should be 82 pennies."

Gebhard Gauss, Carl's father, whipped around and glared at the child. "What are you doing out here?" he snarled. "Get back into that house, now! Move, child!"

Returning to one of the workmen, he tried to control himself: "*Herr* Braun, let me add that again: $34 + 29$: that would be the same as $34 + 30 - 1$ and that would be $64 - 1$ which is 63. Then I need to add 19, so I say $63 + 20 - 1$ and that is $83 - 1$ and that is 82. God in heaven! The child is right! Curses! *Herr* Braun, here is your money—82 pennies."

His voice was still shaking as he turned to his second workman. "*Herr* Schmidt, your total is 27 pennies plus 32 pennies plus 24 pennies and that makes a total of ... um, let me see ... yes, a total of 83 pennies. You'll be back tomorrow? Shortly after dawn? Good." And with that Gebhard went inside.

"Well, that's done, Dorothea," Gauss' father announced as he strode into the kitchen. "Where's my supper? I'm hungry!"

"Gebhard, what's wrong? Was there a problem with the workmen? You seem upset."

"Nothing's wrong," Gebhard answered. "Dorothea, why on earth did you teach the child to add? He's only three years old! He's still a baby!"

"But Gebhard, you know I can't do figures. He asked me some numbers, and I told him the ones I know. That's all."

"Then it's that brother of yours, always hanging around the house and doting on the child! Honestly, Dorothea. I wish you'd tell Friedrich to mind his own business!"

"Gebhard, Carl really seems awfully clever," said Dorothea. "I think he may have figured it out for himself."

"No three-year-old child can teach himself to add. I know this is your brother Friedrich's doing. I know it! Where is my supper, woman?"

"Here it is, Gebhard." Turning to Carl, she said quietly, "Come with me, Carl. We'll go outside."

As Dorothea sat down on the front step where she could watch Carl play, she thought about what Gebhard had said. She remembered, some time ago—was it last spring?—when Carl had surprised her. They were in the kitchen, and she had taken out six small potatoes and put them on the table before peeling them for soup. Carl had crawled up and grabbed two potatoes. "One

Gauss' birthplace, destroyed during World War II

potato, two potato.... *Mutter* [mother], one potato, two potato, then...?" asked Carl.

"What is it, child?" she had asked.

"One potato, two potato, then?"

"Oh, I know what you want," Dorothea had said. "Three potatoes."

"Ya!" yelled Carl. "One potato, two potato, three potato! Whee! Then?"

"Four potato," answered Dorothea slowly.

"One potato, two potato, three potato, four potato, then?" asked Carl.

"Next comes five and then six," said Dorothea, "and that's all the potatoes I know how to count."

"One, two, three, four, five, six," said Carl. "Six, five, four, three, two, one."

As Dorothea thought back to that scene in the kitchen, she was sure of it. She hadn't pushed Carl to learn to count. Carl had asked questions, and she had answered.

Then she thought back to another day maybe a week later when her brother had stopped by. Carl had been ready for him.

"*Onkel* Friedrich, what comes after six?" Carl had asked. "*Mutter* doesn't know."

"Seven," Friedrich had said.

"Then what, *Onkel* Friedrich?" Carl had asked.

Friedrich had told him some more numbers, and when he stopped, Carl had asked, "Is that all?"

"No, but if I tell you once, you won't remember, so let's use the numbers I just told you," said the *Onkel*.

Then Carl had recited the long list. He already knew more numbers than his mother knew! "What comes next, *Onkel* Friedrich?" Carl had asked.

Friedrich had answered once again, and he must have told Carl enough that he could make up more numbers all by himself. He kept right on naming more and more of them.

"*Onkel* Friedrich," Carl had asked, "is there anything after one hundred?"

"Yes, it goes on and on. It never stops," Friedrich had said. Then Friedrich had told him something more, and then Carl had gone on once more by himself. In the end, Carl had started over and counted from the beginning again, and Friedrich had agreed that he had done it just right.

"I didn't forget them," Carl had said. "I told you I wouldn't forget."

"No, you don't seem to forget anything," Friedrich had said.

Dorothea thought back to another afternoon. Once again, she and Carl and her brother had been sitting together on the front step.

"*Onkel* Friedrich," Carl had asked, "can we write the numbers?"

"Oh, yes, Carl," answered Friedrich. "Look." And Friedrich had picked up a stick and drawn a straight line and then a curvy line in the dirt. "Here is one: 1; and here is two: 2."

"Where is 3?" Carl had asked impatiently.

"Here they all are," Friedrich had said. It had looked as if Friedrich was drawing the shapes in some kind of order. Then he had read the list back to Carl. "1, 2, 3, 4, 5, 6, 7, 8, 9." Then he had stopped and corrected himself: "No, Carl, I should have started with this: 0," and he had drawn a circle. "So now the list is complete: 0, 1, 2, 3, 4, 5, 6, 7, 8, 9."

"What is the circle?" Carl had asked.

"It's zero, the number for nothing," *Onkel* Friedrich had said.

"Why do we need zero?" Carl had asked. "Why would you talk about zero potatoes if you don't have any? That's silly!" Dorothea had to agree with her son. Why would anyone need a way to say nothing?

"Well, your mother might look in the pantry and see that she has zero potatoes," Friedrich had said.

"And that would mean she needed to go to the market," Carl had said. "I get it!"

Then Carl had wanted Friedrich to write more numbers, but Friedrich had explained that he didn't need any more marks. Those marks that he had already made in the dirt were all that he would need. Apparently you can write the rest of the numbers using just the squiggles and lines that Friedrich had drawn.

Dorothea was amazed as she thought back to that afternoon. She hadn't followed all that they said, but her brother and Carl had been very excited. Gebhard said Friedrich was being pushy with the child, but Friedrich had just answered Carl's questions. It was Carl who kept on pushing. Carl wanted to know—he demanded to know. She couldn't stop Carl's curiosity, and she didn't think she wanted to even if she could.

Dorothea stopped her daydreaming and looked at Carl. She had been sitting here on the step for quite some time, and he had been busy. He had been running back and forth across the street between the canal and the step where she sat. He had been making a pile of pebbles beside her.

Now, what was he doing with the pebbles? He was putting them in lines. There was one set of five, and another set of five, and then another, and then another. "*Mutter*, there are 20 pebbles: 5, 10, 15, 20. I could also count them by fours. I could say 4, 8, 12, 16, 20. I wonder if I could count them by threes. Do you think I could do that, *Mutter*?" asked Carl.

"Child, I don't know," Dorothea answered in confusion.

"Well, I guess I'll just have to try: 3, 6, 9, 12, 15, 18, and then I have two left over. That's interesting," said Carl. "I didn't have any left over when I counted by five and four, but I have two left over when I count by three. *Mutter*, do you suppose I'll have any left over if I count by twos? Oh, no! Of course I won't. If I can count by fours, I can certainly count by twos because four is two twos. I wonder what would happen if I count by eights."

Sums

The following week Dorothea was standing in line at the butcher's shop when the women heard a commotion outside. When they ran out to look, they saw a man striding toward them carrying a small child.

No, it couldn't be! It was her baby—little Carl—and he was soaking wet!

"*Frau* Gauss, I found your child in the canal—he must have fallen in. You weren't home, so I figured you'd be here and I brought him to you at once. I think he is okay, but you'll have to judge for yourself," the man said.

"Thank you so much, *Herr* Braun!" said Dorothea. "Carl, are you all right?"

"I'm not hurt, *Mutter*, but I'm cold. Can we please go home now?" asked Carl.

"Of course we can. I'll come back later and get the meat," she answered as he snuggled against her shoulder as she carried him home. "What were you doing in the canal, Carl?"

"I was experimenting. I found some leaves, and I dropped them into the canal, and then I watched to see where they moved the fastest. It was very interesting. The ones that I put on the side moved slowly because they kept bumping against the wall. The ones that I put farther out went faster. So I reached down very carefully to push the ones that were near the wall farther out. They moved faster then, and I saw that I needed to move them all away from the wall. I guess I lost my balance. I'm sorry. I didn't mean to fall in, but now I know that if I want something to move along in the water, I need to put it out away from the edge so it doesn't bump against the wall," said Carl. "I'll never make that mistake again."

"It was not a good idea to play in the canal, Carl," scolded his mother gently. "I don't want you to play there. I think it is dangerous. Do you realize how lucky you are that *Herr* Braun saw you?"

"Well, I yelled! He wouldn't have seen me if I hadn't yelled," said Carl.

"That may be, Carl. I'm glad you yelled, but I don't want you playing in the canal again," she said.

"All right, *Mutter*," said Carl. "I'll stay away from the canal unless you are there with me."

"Now we need to put some dry clothes on you. Ach! Those wet clothes smell bad!" she said.

"*Mutter*, why do we have a canal beside the street?" asked Carl.

"I don't know," she admitted. "Well, yes, I do know. That's where I pour the dirty water after I wash the dishes or the clothes. It takes the dirty water away."

"So that's why the water is smelly! What would we do if we didn't have the canal?" asked Carl.

"I'm not sure," Dorothea answered.

"I know! We'd have to dig one. We could use father's big shovel, but then we'd have to put stones on the bottom so it would look pretty," said Carl. "And after we finished, we would be sure to put father's shovel back in the shed so he wouldn't get angry."

"Yes, that's true," said Dorothea.

"I wonder where the water goes," said Carl.

"Maybe it goes to the river," suggested his mother.

"I'll ask *Onkel* Friedrich," said Carl. "He'll know."

"Hello, Dorothea!" Friedrich Benze greeted his sister one afternoon a few days later. Dorothea and Friedrich had always been close, but while Dorothea had spent her childhood at home, learning to sew and cook and clean, Friedrich had been sent to school for a few years before being apprenticed to a weaver in town.

It didn't take long for Friedrich to master the basics of weaving so that he could set out on his own. As a clever young man, Friedrich tinkered with the weaving process to see if he could make some interesting patterns. Then one day he caught sight of a lovely piece of damask in the market. "I wonder how they did that?" he had asked himself, and he began to experiment with raising some of the threads as he threw the shuttle to one side and then raising other threads as he threw the shuttle back again. He realized that if he wanted to make a pattern it would be more efficient if he could use notched cards to move certain threads up and other threads down as the shuttle went back and forth across the warp. It was a mathematical process, similar to the punched cards used in computer programming 200 years later. As he worked, he became more and more skillful, and the resulting fabric was beautiful. It didn't take long for him to earn a reputation as a skilled artisan.

"I just finished a lovely piece of damask for a tablecloth for the Duke," Friedrich began, "and I thought I'd come visit with you and Carl for a bit. Where is that bright little boy?"

"I think he's in the garden," said Dorothea. "Carl! *Onkel* Friedrich is here!"

"I'm coming, *Mutter*!" said Carl. "*Onkel* Friedrich! *Onkel* Friedrich!" shouted Carl as he ran into his uncle's arms.

"Friedrich, tell us about the tablecloth you've made for the Duke," said Dorothea.

"Who's the Duke?" asked Carl.

"The Duke is our prince," explained Friedrich. "He is a very important man. The damask I weave for him must be the very best, and this piece that I have just made is lovely and very large. It's one-and-a-half yards wide and three-and-a-half yards long. I have woven Duke Ferdinand's coat of arms into the pattern along with scenes from a lovely garden."

The Duke's family had been ruling the Duchy of Braunschweig, the part of northern Germany where they lived, for many years. The current Duke was an intelligent, well-educated man who understood the needs of his duchy. Friedrich and Dorothea knew that they were fortunate to live in the Duchy of Braunschweig.

"Is this the first tablecloth you have made for the Duke?" asked Dorothea.

"Oh, no! I've made several for him, but this one is the most beautiful and the biggest. It's the biggest piece I've ever made. He will pay me ten *Thalers* for it," said Friedrich.

"Ten *Thalers*!" gasped Dorothea. "So much?"

"It represents more than a month of hard work, Dorothea," said Friedrich. "The Duke knows it will be beautiful, and he does not take advantage of me. I will give him good value for his *Thalers*."

"The Duke must be very rich!" said Carl quietly.

"Yes, the Duke is rich, but he has many responsibilities," said Friedrich. "Our duchy is well run. He takes good care of us. I like to think of him enjoying my weaving."

"Friedrich, could you bring the tablecloth here so that I can see it before you give it to the Duke?" asked Dorothea.

"I'd be glad to do that," said Friedrich.

"Do you have more orders for weaving?" asked Dorothea.

"Yes, there are two young women who each want several pieces before they get married next month. I have plenty to do," said Friedrich. "My problem is that there are not enough daylight hours."

Carl asked, "If the Duke pays you ten *Thalers* for that beautiful tablecloth and then you make him three more beautiful tablecloths, that would be a total of four tablecloths. Would you get four times ten *Thalers* or 40 *Thalers* in all?"

"That's right, Carl," said the *Onkel*. "How much would I get for ten tablecloths?"

"You would get 100 *Thalers*!" said Carl. "And for 12 tablecloths you would get 120 *Thalers*! You are going to be very rich, *Onkel* Friedrich!"

"How many tablecloths did the Duke order, Carl?" asked Friedrich.

"Only one," said Carl quietly.

"That's right. I will get just ten *Thalers*," said Friedrich.

Changing the subject, Carl asked, "*Onkel* Friedrich, where does the water in our canal go?"

"Where does it go? Well, let me think. First it flows into the Schunter River. Then it flows into the Weser, a much bigger river, and then eventually it flows into the North Sea."

"So does that mean if I drop a stick in the canal that it will reach the sea tomorrow?" asked Carl.

"I doubt that it would happen that fast. The water doesn't move very fast, and you know there is a possibility that it would get caught by a rock or a log," said Friedrich.

"If it did, would it be stuck there forever? Wouldn't it ever get to the sea?" asked Carl.

"Carl, I really don't know. I think it is possible that it could reach the sea, but it is also possible that it would not reach the sea," said Friedrich.

"And what if another boy picked it up and put it in his pocket?" asked Carl.

"That is another possibility," said his *Onkel*.

"Or what if an older boy picked it up and took it to his mother and she put it into the fire in the stove so she could cook his dinner?" asked Carl.

"Yes, that is possible too, although it would have to dry out first," said Friedrich. "Wet wood does not burn well."

"I'm going to drop a stick in the canal right now," Carl announced. "Don't worry, *Mutter*. I won't get too close to the edge. Maybe it will go all the way to the North Sea. I'd like to go to the sea some day. Have you ever been to the sea, *Onkel* Friedrich?" asked Carl.

"No, I haven't," said Friedrich, "but I have seen pictures of it."

"Could you show me a picture of it?" asked Carl.

"I'll see if I can find one for you," Friedrich promised.

That evening, Friedrich brought the Duke's tablecloth to show his sister.

"Oh, Friedrich," Dorothea said, "this is beautiful! And it is so big!"

"Yes, the messenger told me that the Duke's table is three yards long," Friedrich said. "I don't think I have ever seen a table that big."

"*Onkel* Friedrich," Carl began, "may I look at the tablecloth?"

"Of course you may, Carl," Friedrich said.

"So the way you make the picture," Carl asked, "is that some of the threads go up and down every other row, but then other threads go on top for quite awhile before they go back down. That makes those parts look shiny while the ones where the thread goes up and down close together don't look so shiny."

"That's right, Carl," said Friedrich.

"But *Onkel* Friedrich," Carl continued, "how did you know which threads to move which way?"

"Well, I had to make a detailed drawing of the Duke's seal and of the garden, and then I had to decide where I wanted it to be shiny and where I wanted it to be dull so that it would look like the picture. Then I had to figure out the sequence of ups and downs in each row."

"Does that mean," Carl continued, "that you could weave any picture you wanted into the damask?"

"Yes, Carl, I suppose it does, but I would never choose something ugly to weave. Damask is supposed to be beautiful."

"Well, this is beautiful," Carl said.

It was the end of the day a few weeks later, and Carl had gone out to his father's work area to see what was happening. "Father, does this mean that you have 15 stones to work on?" asked Carl.

"Why do you ask me that?" snapped Gebhard.

"It says on the slate that you had three, and then you got five more, and then you got seven more. So that would make 15 in all," said Carl.

"How do you know that?" asked his father.

"I am right. I added them. You have 15," said Carl.

"Yes, you are right," said Gebhard. "Tell me what this is." He wrote $36 + 17$ on a slate.

"Well, that would be $36 + 20 - 3$ and that would be $56 - 3$ which would be 53," answered Carl.

"Carl, can you read this number?" asked Gebhard.

"That's six hundred twenty-nine. Why?" asked Carl.

"Good Lord, child," said Gebhard. "Maybe you are different from other children."

Gebhard stopped and thought. What was he going to do with this child? Suddenly an amazing idea came to him. He could invite the neighbors in and impress them with his clever little boy! No one would expect a three-year-old child to read or add numbers.

Sunday afternoon, as the neighbors sat in the family's living room, Gebhard wrote an addition problem on a slate. Carl sat on his mother's lap across the room.

"Carl, add these numbers. What is their sum?" Gebhard demanded.

Silence.

"Carl, what is the sum of these numbers?" Gebhard shouted. Gebhard wondered what was the matter with the child. Why wouldn't he perform?

The neighbors were not sympathetic. They had known all along that it was just a trick. No three-year-old child could add those numbers! Why, many of the neighbors couldn't add them either! They left, whispering among themselves. "Wouldn't you know *Herr* Gauss would put the child up to something like that? That child couldn't add if his life depended on it!"

"*Mutter*, why did the neighbors come here today?" asked Carl after everyone had left.

"Because your father wanted to show them that you can add, Carl," answered his mother. "He is very angry. Why didn't you do it for him?"

"Do what?" asked Carl. "Were there some numbers I was supposed to add?"

"Didn't you see what your father wrote on the slate?" asked Dorothea.

"No. I knew he wrote something, but I couldn't see it, and I knew he wouldn't like it if I ran over to look at it," said Carl.

"No, I guess he wouldn't have liked it. Go get the slate, Carl, and tell me what it says," said Dorothea.

"Here it is, *Mutter*!" said Carl. "It says 37 + 51."

"So what is the answer?" asked Dorothea.

"Well 37 + 50 is 87, so this should be one more than 87, so that would make it 88," said Carl.

"Amazing!" she gasped.

"I am right. You know that I am right, don't you?" he asked.

"Child, I don't know! I don't know how to read or add."

"Well, I am right. That is the answer. If Father asks you, the total is 88."

The Duke of Braunschweig

One afternoon Carl and his mother were walking the two blocks to the market in town. It had rained all night and all morning too, clearing only around noon, and the muddy road still had many puddles.

Suddenly there was a commotion, and a large, elegant coach pulled by four horses rumbled past them. "Oh *Mutter*! What was that? We got splashed!" yelled four-year old Carl.

"I'm sorry, child. I didn't see him coming," said Dorothea.

"See who coming?" asked Carl.

"That was the Duke in his carriage," she explained.

"Is that the same duke that *Onkel* Friedrich made the tablecloth for?" asked Carl.

"Yes, it is."

"But I thought *Onkel* Friedrich said he is nice. This wasn't nice! He almost ran us over, and we are all muddy and wet," said Carl.

"That's all right, Carl. It's turned into a nice warm day. We'll dry quickly," she said.

"But why did he do it to us?" he asked.

"He didn't mean to do it to us, Carl. The Duke is a very busy man, and he always has to move quickly. He didn't even see us," she explained.

"But he was standing on the back of the carriage! He must have seen us," said Carl.

"No, Carl, that wasn't the Duke. That was his footman—his servant. The Duke was inside the carriage," she said.

"So you think the Duke doesn't even know that he splashed us?" asked Carl.

"Probably not," she answered. "He is a very important man. He is a prince."

"I think when I grow up I would like to be a prince," said Carl. "I would like to ride in a carriage like that."

"No, Carl, that won't happen. We are simple folk. We work hard and we do the things we need to do, but we will never have much money," said Dorothea.

"When I grow up, I will be rich. I will buy you a new straw hat every spring, *Mutter*," said Carl.

"I doubt that you will be that rich, but I suppose you might. A new hat every couple of years would be enough. I certainly don't need one every year. However, whether or not you are rich, you will not be a prince. The Duke's son will be the next duke. The only way you can be a duke is if your father was a duke before you, and your father is no duke!"

"No, Father has to walk to work, but he is a busy man too," protested Carl.

"Yes, he is busy, but in different ways than the Duke is busy."

"Where is the Duke going?" asked Carl.

"I don't know, child. He has many things that he needs to do. He is in charge of Braunschweig. He's our ruler," explained his mother.

"Well, then, what does he do?" asked Carl.

"I don't know. He is very rich and very important, and it is up to us to stay out of his way."

"I'd like to meet him someday. Do you suppose I could?"

"I doubt it, Carl. He works with the important people in town. It is up to us to keep our little part of the town going well."

"Is the Duke wise?" asked Carl.

"I hope so. We depend on him to be wise," said Dorothea.

The Schoolroom

When Carl was seven years old, he started going to the Katherineum—the local elementary school. He walked the two blocks to school, talking with other boys and playing with things they found along the way. Only boys attended *Herr* Büttner's school, at least 100 of them. The girls, like Carl's mother before them, didn't go to school. If a girl was taught, she was taught by her mother or a tutor at home. Because it was assumed that she would spend her life sewing and cooking and bringing up the children, most people couldn't see any reason for a girl to learn to read. However, everyone agreed that boys needed some formal schooling, and it was time for Carl to get his.

The schoolroom had a low ceiling and an uneven floor, making it feel close and cramped. The teacher, *Herr* Büttner, was a barrel-chested man with a small beard on his chin. He never cracked a smile as he paced across the room. He carried a small whip in his left hand, using it vigorously and often as he kept all those squirming boys under control.

Much of their time in school was spent waiting—waiting for others to answer a question and waiting nervously for their own turn to be called on to recite. For most of the children, it was nerve-racking. They knew that they could be called on at any moment and that the chances were good that they would not know the answer to their question. The whip was a constant threat, ready to correct them whenever they made a mistake. Even if they knew the answer, they were often too scared to get the words out.

However, Carl was not intimidated. He already knew most of what *Herr* Büttner was teaching them. What Carl didn't already know he learned easily and quickly. No one had to tell him anything twice. His quiet confidence was sometimes interpreted as arrogance when he was older, but to Carl it was natural. He knew, and he knew that he knew.

"All right, Müller, tell me the principal parts of the verb *to eat*."

"I-I-I-I don't know, sir."

"You don't know the principal parts of the verb *to eat*, Müller? Are you hungry?"

"No, sir. I et my breakfast before I came to school."

"You et your breakfast, eh? (Thwack! Thwack!) Do you mean you *ate* your breakfast? Perhaps if I remind you more forcefully you will remember next time. (Thwack!)"

"Gauss! What are the principal parts of the verb *to eat*?"

"*Eat, ate, eaten*, sir, are the principal parts of the verb *to eat*, sir."

"Very good, Gauss. Now, Müller, what did Gauss say are the principal parts of the verb to eat?"

"*Eat, et, eaten*."

"Müller, you need to listen more carefully. (Thwack!) Koch, what are the principal parts of the verb *to eat*?"

"*Eat, ate, eaten*, sir."
"All right. Koch, are you hungry?"
"No, sir, I always eat breakfast before school."
"Yes, Koch, but what did you eat for breakfast today?"
"I et bread and cheese for breakfast today."
"You *et* it, did you? (Thwack! Thwack!) Are you sure you shouldn't have said 'I *ate* it'?"
"Yes, sir. I ate bread and cheese for breakfast today."
"Good, Koch. And Müller, what did you eat for breakfast today?"
"I have ate bread for breakfast today, *Herr* Büttner."
"You have *ate* it, Müller? Are you sure you shouldn't say *I have eaten it*? Are you sure, Müller (thwack!)?"
"Yes, sir. I have eaten bread for breakfast today."
"Stop your sniveling, Müller. What are the principal parts of the verb *to eat*, Müller?"
"The principal parts (sniff, sniff) of the verb (sniff) *to eat* ... are *eat, ate, ... eaten, Herr* Büttner."
"Well, Müller, you finally got it. Make sure you don't forget it."

Büttner's assistant, 17-year-old Martin Bartels, selected, cut, and sharpened the goose quills that the children wrote with and helped them with their writing. Penmanship was an important part of the basic education of the day, and, although writing with a quill is difficult, Carl mastered it quickly and well.

"Schiller, what is the principal city in the Duchy of Braunschweig?"

"Braunschweig is the principal city in the Duchy of Braunschweig, sir."

"Gauss, what are the other major cities in the Duchy of Braunschweig?"

"Goslar, Helmstedt, and Wolfenbüttel are the other major cities in the Duchy of Braunschweig, sir."

"Müller, what are the four major cities of the Duchy of Braunschweig?"

"The major cities of the Duchy of Braunschweig are Goslar, Helmstedt, and Wolfenbüttel, sir," Müller answered.

"Aren't you forgetting something, Müller? Isn't there another major city in the Duchy of Braunschweig, Müller?" asked *Herr* Büttner.

"No, sir. I don't think so, sir."

"No, Müller, you don't think at all. (Thwack!) Koch, what is the principal city in the Duchy of Braunschweig?"

"Braunschweig is the principal city in the Duchy of Braunschweig, sir."

"Did you hear that, Müller? (Thwack!) Can you name the four major cities in the Duchy of Braunschweig now, Müller?"

That evening after supper, Friedrich Benze stopped in for a visit. "*Onkel* Friedrich," Carl asked him, "what does the Duke do?"

"Why do you ask that, Carl?"

"Well, we were learning in school today about the Duchy of Braunschweig and the major cities in the Duchy. Is the Duke in charge of the whole Duchy?"

"Yes, he is our prince. He has to keep track of what is going on here and in Helmstedt and in Wolfenbüttel and in Goslar and in all the places in between. When there is trouble, he has to step in. He is a very intelligent man and he seems to keep things running smoothly. We are lucky to have him as our duke."

"How come we never see him? If he is so important and he keeps track of all that is going on in the Duchy, he should be out in the market, talking to people, shouldn't he?"

"Carl, he doesn't need to see it all for himself. He has many people working for him who report back to him on all that is happening. He can't be everywhere all at once."

"But he knows you. Remember when you wove that beautiful tablecloth for him when I was little?" asked Carl.

"Yes, I remember," said Friedrich. "But the Duke doesn't worry about a little thing like a tablecloth."

"But it was big! And he talks to you sometimes, doesn't he?" asked Carl.

"No, Carl, I don't think he knows who I am," said Friedrich.

"Well, he should! I would like to meet him, and then I could introduce him to you. Perhaps I could advise him on something," said Carl.

"You will have to become much wiser than you are now if you want that to happen, Carl. I have never met him, and your mother and father haven't either. But I hear that he is a kind man and I know that he is an excellent administrator."

"What's an administrator, *Onkel* Friedrich?"

"Someone who runs things. He is someone who organizes people—who tells people what to do so that all the important things get done and everyone is relatively happy."

"You know, *Onkel* Friedrich," Carl said, "his carriage splashed *Mutter* and me one time when we were going to the market when I was little. It was a very grand carriage, and we couldn't see him. We saw only the footman on the back. It was moving very fast. We had to stand back so that we wouldn't be run over by one of the wheels."

"Yes, Carl, the Duke doesn't have time to walk from place to place the way we do. He sometimes has to visit Goslar or Helmstedt on the same day that he also has something important going on in Braunschweig."

"I know where the palace is. It looks very big. Is the Duke the only person who lives there?"

"No, no, Carl. He has a large staff of people who work for him, taking care of the palace and preparing food, and other people who do the work of running the Duchy."

"But what do they do, *Onkel* Friedrich? Do they make lists on paper? Do they sharpen his quills so he can write more?"

"Carl, there are many questions that I can answer, but that isn't one of them. I really don't know what the Duke or his staff do, but I know they do it well because everything seems to run pretty well. I think you will have to wait until you are older to get the answers to those questions."

"He is very rich too, isn't he?" asked Carl.

"Yes, I believe he is very rich," said Friedrich.

"I plan to be rich someday, too," said Carl, "but *Mutter* says I can't be a prince."

Each afternoon when Carl got home from school, he had work to do. Since his family was poor, everyone had to work.

Carl's chore was spinning flax to make linen thread. It was a long, boring job, but Carl didn't mind. After all, nobody bothered him as he spun, and he could spin all the arithmetic he liked in his head as he worked.

"Let's see. I know that $8 \times 8 = 64$, that $7 \times 7 = 49$ and that $7 \times 8 = 56$. There must be a number between 7 and 8 that you could multiply times itself to get 56. If I multiply $7\frac{1}{2}$ times itself (that's $7\frac{1}{2} \times 7\frac{1}{2}$), that's $\frac{15}{2} \times \frac{15}{2}$ which is $\frac{225}{4}$ which gives me ... hmm ... $56\frac{1}{4}$. That's close, but it's too big. That's funny. I would have thought that the average of 7 and 8 would work. Maybe the number I'm looking for is another kind of average. I'm sure there is a number that I could multiply times itself and get 56. I'll bet that for every number there is a number that you can multiply times itself to get that as an answer. It ought to have a name. Maybe it already has one. I wonder. I must ask *Onkel* Friedrich."

So the hours at the spinning wheel were not as tedious for Carl as they might have been, but it was a good thing he could do all that calculating in his head. If his father had known about his arithmetic on the sly, he would have exploded in a fit of rage.

On the afternoon of Christmas Eve, Carl and his family were invited to a Christmas celebration at the home of Carl's godfather, Georg Karl Ritter. Carl's mother Dorothea had worked for the Ritter family before her marriage to Gebhard, and when her son was born, *Herr* Ritter had been delighted to become Carl's godfather. The Ritters' home was more comfortable than the

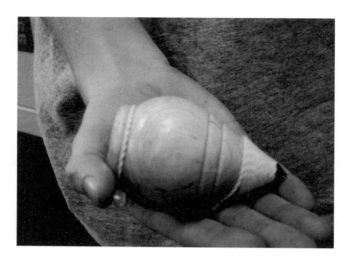

Top such as Gauss might have received at Christmas from his godfather

Gausses', and when the sun went down, *Herr* Ritter lit an oil lamp so his guests could see to play games and sing together. Carl was astonished. Although it was dark outside, inside the cozy house it was light enough to see the other people around him. "What a magnificent invention!" Carl thought. "When I grow up, I would like to have a lamp for my house."

When Carl opened his present later in the evening, he was amazed to find a top. *Herr* Ritter showed him how to wind the string carefully around the top and then, with the top standing on its point, to pull the string so that the top would spin on its own. Carl had never seen anything like it. What kept the top from falling over? *Herr* Ritter was delighted that he had been able to procure such a fine gift for his young godson.

Arithmetic

After Carl's first two years of school, it was time to begin arithmetic. One day when *Herr* Büttner wanted to keep the boys quiet for awhile, he gave them an assignment he knew would accomplish that. He asked them to find the sum of the first 100 counting numbers: $1 + 2 + 3 + 4 + 5 + 6 + 7 + \ldots + 96 + 97 + 98 + 99 + 100$. The other boys quickly set to work on their enormous chore, adding $1 + 2 = 3$; $3 + 3 = 6$; $6 + 4 = 10$; $10 + 5 = 15$; $15 + 6 = 21, \ldots$. This was going to take them a long time, but *Herr* Büttner's whip was ready to straighten out any boy who gave up on the job.

Carl used a different approach. Rather than start in on the adding immediately, he sat and thought a minute. Then he wrote the answer on his slate and walked to the front of the room to put his slate on *Herr* Büttner's table as the first one in the pile. As the other boys finished, they would put their slates on top of Carl's in the order in which they finished. *Herr* Büttner looked at Carl's slate, saw just one number, and glared at Carl. What a pleasure it would be to correct that child when it was time to check the answers! In the meantime, Carl sat at his desk and waited patiently for the others to finish. Carl knew that his solution was correct.

If he had been easily intimidated, *Herr* Büttner's glances would have made him tremble. However, Gauss' confidence in his own reasoning then and in later years was unshakable. It was time to move on to the next challenge.

Carl's thoughts were far away. "Let's see. $13 \times 13 = 169$. If I multiply the number one less than 13 times the number one more than 13 (that's 12×14), I get 168. That's one less than 169. Okay. What happens with 14×14? $14 \times 14 = 196$. If I multiply one less than 14 times one more than 14 (that's 13 times 15), I get 195. That's one less than 196. Will it always work? I'll bet it will. Let's see. $25 \times 25 = 625$. 24×26 is—let's see—yes, it's 624, and that is one less than 625. Yes. It looks good. I'll bet the answer is always going to be one less than the square of the number in between."

An hour later, when all the slates were finally stacked and ready to be checked, Büttner reached into his pocket for the slip of paper with the answer on it, and then he began to check the slates. One slate after another had the wrong answer. Some boys had made their first mistake early in their calculations, so naturally they were doomed to failure almost from the beginning, and most of their hard work was completely wasted. Some waited until close to the end to make their first mistakes, but even so their answers were wrong. This assignment was proving to be a trial for the boys. The whip got plenty of use.

Finally *Herr* Büttner reached Carl's slate at the bottom of the pile, and there he found the correct answer: 5,050. How had Carl gotten it? He had spent almost no time on it, he hadn't done any figuring on his slate, and it looked suspicious.

Arithmetic

He demanded an explanation. "Tell me, boy, how you got this answer!"

Carl stood up and began to speak. "Well, sir, I thought about it. I realized that those numbers were all in a row, that they were consecutive, so I figured there must be some pattern. So I added the first number and the last number: $1 + 100 = 101$. Then I added the second and the next to last numbers: $2 + 99 = 101$. It started to make sense. $3 + 98 = 101$; $4 + 97 = 101$; $5 + 96 = 101$. If I continued adding pairs of numbers like that, I would eventually reach $50 + 51$. That meant I would find 50 pairs of numbers that always add up to 101, so the whole sum must be $50 \times 101 = 5,050$. There was no need to add up all the numbers, sir."

Herr Büttner was dumbfounded. He had learned a formula to figure out problems like this. After all, he certainly hadn't sat down and added up all those numbers! But how could this ten-year-old boy, son of a common workman, figure it out? It had taken *Herr* Büttner many years of schooling to master techniques like that, and now young Gauss had figured it out for himself in just a few minutes! Perhaps he had underestimated the boy.

Büttner ordered a more advanced arithmetic text and assigned Martin Bartels, his assistant, to help Gauss work through it. However, it soon became clear that Carl had already figured out all the concepts in that book, and that he needed something even more advanced. Fortunately for Carl, Martin Bartels was also fascinated by mathematics (he later became a professor of mathematics at a major university in Russia), and together they worked through a book on elementary algebra. Martin was only seven years older than Carl, and they rapidly became close friends, working together to figure out mathematics as they went along.

A few months later, *Herr* Büttner decided it was time to talk with Carl's father about his extraordinary son. *Herr* Büttner didn't like calling on parents at home, but occasionally there was no alternative. He knocked at the door, and waited. When Gebhard threw open the door, he was clearly shocked to find the teacher.

"Carl! Get out here! Your teacher is ... "

"Please, *Herr* Gauss, I'd prefer to talk with you alone."

"I always knew it would come to this. That boy! I knew he was headed for trouble as soon as that uncle of his started making so much of him. He thinks he's so smart."

"No, sir, that is not a problem."

"It isn't? Well, then, what is it? Has he been misbehaving in school? Smart-mouthing? Mocking the other children? Bullying them?"

"No, sir. If you would just listen a minute. Carl is always polite, and he does his work well. The situation is this: I have already taught Carl all that is in the normal curriculum. No. That's not right. Let me say it another way: Carl already knows all that I normally teach the boys. He has already figured it out on his own. You see, *Herr* Gauss, your boy is a genius. He is destined for great things—very great things. He will be famous some day. You will be proud of him. I came to talk with you, sir,

Arithmetic

about giving him what he needs at this point, so that he can go on and accomplish the great things that should be ahead of him."

Gebhard was stunned. This was not what he had expected. He had no idea how to deal with this. He had always assumed that his younger son was destined for a life just like his own—a difficult life of rough, backbreaking labor. And he had no extra money for a fancy education or expensive books.

Herr Büttner went on: "Martin Bartels, my assistant, is already spending extra time with your boy. The two of them are working their way through algebra, but Carl wants to devote more time to it so that he can get the foundation necessary for the real work he wants to do. Would it be possible for your boy to work in the afternoons and evenings with Martin so that they can make the progress they both want?"

"I'm sorry, *Herr* Büttner. Carl doesn't have time for extra schoolwork. Like every other boy in this town, he has work to do after school. We expect him to spin flax. A growing child eats a great deal and it is only fair for him to contribute his fair share to the family. That's what I did when I was his age, that's what his older half-brother Georg does, and that's what Carl will continue to do."

As the discussion continued, the teacher finally convinced Gebhard that his son Carl was a genius and that he needed the time to develop his remarkable talents. Büttner commented that in 20 years of teaching he had never encountered the likes of Carl. Gebhard reluctantly agreed to excuse Carl from spinning so that he could study after school.

Carl could hardly believe his good fortune. From that time on, Carl and Martin spent many afternoons and evenings happily studying geometry and algebra as well as Latin together.

"But, *Herr* Büttner, how will I ever find the money to pay for all this education? I am a simple, hard-working man. There is no money left over at the end of the week."

"I believe we can work something out. We will make arrangements for the Duke or someone else to sponsor him. I don't think you need to worry about the money."

Carl's biggest problem now was that in the winter the evenings were dark, with the sun going down before four o'clock in that northern latitude. The family did not have the money to buy candles or lamps, so they simply went to bed when it got dark.

Carl was frustrated. There was no doubt that it was dark outside and dark in the house, and dark everywhere else. Reading was not an option, and there was a limit to the effective work he could do in his head. Then as he walked around outside one Saturday afternoon he noticed a turnip lying on the ground. "I wonder if I could make myself a lamp," he mused. He took an axe and chopped the turnip in half. Then he gouged out the middle of it. He went to the kitchen where his mother was working. As always, she was eager to help him in whatever he wanted to do.

"*Mutter*, do you have some fat that I could use?"

"Yes, here are some drippings from Sunday's roast. What do you want it for?"

Arithmetic

Turnip lamp, made from a turnip cut in half and hollowed out. The wick is a scrap of cloth.

"Oh, nothing. Do you have a piece of cloth that I could use—just an old scrap?"

"Would this do? It's just a bit left from that shirt you wore out. I've used the rest of it for patches, but this piece from the elbow was too worn."

"That will be great. Thanks, *Mutter*." Then Carl put the cloth in the fat leaving one greasy corner sticking up. "Now, can I get just a bit of fire from the stove?" As he said that, he took a splinter of wood, lit one end of it in the fire, and used it to light the rag in his "lamp." It burned. It flickered. It burned again. It wasn't the beautiful clear light that you would get from a proper oil lamp or a candle, but it was light. Carl was delighted.

His mother looked at him with amazement. "What are you going to do with that?"

"I'm going to read upstairs after supper when everyone else is asleep. I have too much work to do."

"Carl, I don't think I want you to have a fire upstairs. I think that might be dangerous. Couldn't you use it down here at the table? That would be more convenient anyway, don't you think? You could bring your blanket down with you so you wouldn't get cold."

From then on, his crude lamp with its tiny, flickering flame allowed Carl to work far into the evening, stopping only when he was too tired or too cold to continue. His mother was happy to provide him with more fat as he needed it.

Early Mathematics

In order to master Euclid's *Elements*, the Greek geometry book from 300 BC that was the foundation of most geometry courses for the next 2,000 years, Carl and Martin had to work through the arguments step by step. They had to grasp the postulates (the assumptions that couldn't be proven) and see how to put together the proofs, step by step, from those postulates. The first four postulates were easy enough:

1. Only one straight line can be drawn between any two points.

2. A straight line can be stretched out forever in both directions.

3. If you take any point, a circle of any radius can be drawn using it as the center.

4. All right (90°) angles are equal.

When they came to the parallel postulate—through any point that is not on a line, it is possible to construct only one line parallel to the original line—like so many mathematicians before and after him, Carl wanted to prove it using the first four postulates. It seemed unnecessary to have this separate postulate. The beauty of Euclidean geometry is that it assumes so very few things, and from those assumptions everything else can be proven. He wondered out loud if it would be possible to derive the parallel postulate—if he could prove it using the other four postulates.

"Carl, it can't be done. We need to move on," said Martin.

"Martin, I know there is more to this than we see right now. You're right. We need to move on, but I will return to this later."

Carl and Martin also studied Leonhard Euler's [pronounced *Oiler*] famous algebra textbook. Euler had died when Gauss was a young child, but his writings have formed the backbone of mathematics ever since. His algebra textbook was the bible of algebra. If Carl and Martin wanted to learn algebra, they needed to begin with Euler.

"Well, Carl, we've got to work through the fundamental theorem of algebra. Do you see that it has to do with equations that have only one unknown—either an x or a y, but not both." After all, Bartels was supposed to be the teacher.

"Yes, Martin, I understand that. What it's telling us is that for that kind of equation, the number of solutions is always equal to the highest exponent on the variable in the equation and we are guaranteed that there is a solution, right?"

"Yes, Carl, that's what it says. So that means that this equation, $x + 2 = 5$, has only one solution, $x = 3$, but this equation,

Early Mathematics

$x^2 + 3x = (-2)$, has two solutions because its highest exponent is two. Do you know where those two solutions come from?" Martin asked.

"Yes," said Carl. "First you rewrite it as $x^2 + 3x + 2 = 0$. Then you factor $x^2 + 3x + 2$ into the multiplication of two binomials. You are able to say that if *this* times *that* equals zero, then it must be true that either *this* or *that* is zero. Right?"

"That's right, Carl," said Martin.

"Martin, why do you think an equation like

$$x^2 + 3x + 2 = 0$$

has only two solutions? Why couldn't it have more than two solutions? Or why can't it have only one solution?" asked Carl.

"Listen, Carl. Euler proved it and d'Alembert proved it. I think you have to accept what they said. You can't reinvent all of mathematics."

"Well, I'm not convinced, but I guess I can't do anything about it now," said Carl. "I'll come back to it later."

Another area of mathematics that Carl and Martin studied at this time was infinite series, a topic that all the ancient Greeks (except Zeno with his paradoxes) and most mathematicians up until Newton and Leibniz had avoided.

Carl began, "Let's look at Zeno's race between the hare and the tortoise, Martin, where the hare runs twice as fast as the tortoise. Since the tortoise had a head start, each time the rabbit tries to catch up and reaches the place where the tortoise had been before, he is able to close the gap by only one-half, and by that time the tortoise has already moved on again. This means that the rabbit will never catch up with the tortoise. He will always have half of the remaining distance between him and the tortoise left to run."

"But we all know that the rabbit will win eventually. Where is the contradiction?" Martin asked, remembering his role as a teacher.

"Give me a minute, Martin," said Carl, who realized that infinite series with finite sums would come into play. "You know that if we add up the counting numbers

$$1 + 2 + 3 + 4 + 5 + 6 + 7 + \ldots$$

the answer is infinitely large, but there are other infinite series that add to a finite sum. Let's think about what happens if we add the reciprocals of the positive powers of two:

$$\frac{1}{2^1} + \frac{1}{2^2} + \frac{1}{2^3} + \frac{1}{2^4} + \frac{1}{2^5} + \ldots$$

They have a finite sum, the number 1!"

"How can we be sure of that?" Martin asked.

"Let's imagine cutting a piece of paper in half," Carl continued. "We'll reserve one-half, but the second half we will cut in half again. We'll reserve one of those halves (really quarters of the original paper) and we'll cut the other half in half again. If we keep on cutting and reserving half and then cutting again, the reserved part of the paper keeps growing. Each time, we reserve a

Early Mathematics

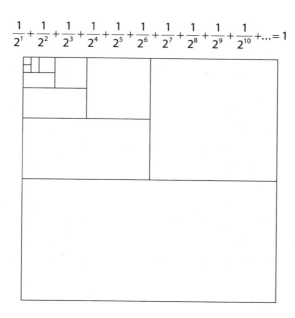

Drawing of the sum of the reciprocals of the powers of two, illustrating the result when you cut a piece of paper in half, and in half again, etc.

smaller and smaller piece as we keep cutting smaller and smaller pieces in half. If we could keep doing this forever, eventually we would have the entire paper laid out on the table in front of us, cut into infinitely many pieces, each piece exactly half of the piece that came before. The sum has to equal one—the original piece of paper that we started with," Carl announced.

"How does that explain the contradiction in the race between the hare and the tortoise?" Martin asked.

"You see, Zeno said that the hare closed the gap by half over and over again. His mistake was in looking at only a finite stretch of the race within a finite time. If he has enough time, the hare will leave the tortoise in the dust!" Carl explained.

"Whoa, I think you have solved Zeno's famous paradox—something he could not explain because the idea of converging infinite series was not even considered in his time."

As Carl and Martin worked together, their roles were beginning to change. At first, Martin had already been familiar with the material, and he had acted as the teacher. As the material became more advanced, the two boys were almost on equal footing since Martin had forgotten some of his more advanced mathematics that he didn't use in *Herr* Büttner's schoolroom. Then, Carl's mathematical genius began to give him the edge. His remarkable ability in mental computation in combination with his phenomenal memory allowed him to do calculations in his head that Martin had to write out. He was able to follow his reasoning further and as a result he saw patterns that were beyond Martin's view. While Martin checked out one example, Carl was able to pursue several of them and put them together to get the big picture quickly. Since there were no calculators at the time, Carl's advantage was even greater than it would be today.

When Carl and Martin talked about the series

$$1 + \frac{1}{2} + \frac{1}{3} + \frac{1}{4} + \frac{1}{5} + \frac{1}{6} + \ldots$$

(the sum of the reciprocals of the natural numbers), it looked to Martin as if it ought to have a limit, but Carl could see at once that the series would sum to infinity if he could go far enough.

Early Mathematics

"Look, Martin. Let's break it into pieces like this:

$$\left(\frac{1}{2} + \frac{1}{3} + \frac{1}{4}\right) + \left(\frac{1}{5} + \frac{1}{6} + \frac{1}{7} + \frac{1}{8} + \frac{1}{9} + \frac{1}{10} + \frac{1}{11} + \frac{1}{12}\right) + \left(\frac{1}{13} + ...\right)$$

Can you see that each of the sets of fractions in parentheses adds up to more than one? If you have an infinite number of these sets of fractions inside parentheses, you will eventually reach infinity. It will take a long, long time, and it is true that each set in parentheses as you go out to the right has more and more terms, and eventually you will reach a set that has an infinite number of terms. With infinity, there is room for many sets and many terms!"

Carl was right. Martin had to admit it.

Martin and Carl had a warm friendship and they had great respect for each other, but early in his life Gauss realized that he would probably never know anyone who could think as fast and as creatively as he did. He might describe his discoveries to others, but he would never find a peer. He could guess that something might be true, run through several calculations to see if the pattern worked, refine his original guess, try it out again, and reach dramatic conclusions in ways that didn't occur to others. As Henri Amiel observed in his journal in the 1870s, talent is doing easily what others find difficult; genius is doing easily what others find impossible.

Part II

The Duke's Protégé
1788–1798

Royal Patronage

The Duke had founded the Collegium Carolinum in Braunschweig to provide a more practical alternative to a traditional university. The college courses provided more than the basic reading, writing, and arithmetic of the elementary school and *Gymnasium* (the high school), preparing students in such fields as engineering, business, statistics, mathematics, and physics. If Braunschweig was to be a major power in the emerging German-speaking world, the Duke knew that education was the key. By the end of the twentieth century, the College had become a technical university, continuing and expanding on the Duke's original idea.

Martin's family had known Zimmermann, the mathematics professor at the college, for many years. They had always planned for Martin to attend the Collegium Carolinum, but they had decided that it was best for him get a little older before studying there. One afternoon, Martin walked over to the college to talk with Professor Zimmermann. "Professor Zimmermann, excuse me for interrupting you," began Martin, "could I take just a few minutes of your time?"

Collegium Carolinum

"*Herr* Bartels! How nice to see you. Yes, please come in. Sit down, sit down. I am always delighted to see you. How are you doing with *Herr* Büttner in the schoolroom?"

"I'm doing very well, thank you, *Herr* Professor."

"Excellent! I'm glad to hear it. And what about your plans to study mathematics? Working with *Herr* Büttner is good experience for you, but I think you should have more ambitious plans," said Professor Zimmermann.

"Yes, sir, I want to continue my studies at the college, but my parents and I don't feel that I am ready yet. However, I have come to talk to you about one of *Herr* Büttner's pupils, a child named Carl Gauss."

"Gauss? I don't recognize the name," said Professor Zimmermann.

"No, sir, he comes from a humble family. His father is a stonemason. He would probably never have crossed paths with you. But it is not the father I want to talk about. It is his son."

"Oh? Is there something special about the son?" asked Zimmermann with some interest.

"Yes, *Herr* Professor. Young Carl is what I think you would call a child prodigy. He figured out all of basic arithmetic on his own before he ever started school. He plays with numbers the way that other children play with toys. His mind is never idle," Martin added.

"That is interesting," said Zimmermann.

"But there is more, sir," continued Martin. "Young Gauss has to hear or see something only once and he knows it forever. When he learns something new, he combines it with what he already knows to produce a deeper and broader understanding. When I mentioned in passing that adding the first 13 odd whole numbers gives a total of 13^2, he took just a few minutes to figure out how to get the sum of the first 13 even numbers. Then he came back to me the next day with a way to get the sum of the first 26 whole numbers. Then he went on to generalize a formula for figuring the sum of the first n whole numbers, no matter what number n might be. He is simply amazing! He has already mastered algebra and geometry. *Herr* Büttner and I truly believe that he is a genius. Is there any way that I could arrange for you to meet him?"

"He certainly sounds remarkable. I would like very much to meet this *Wunderkind*! How about next Tuesday afternoon? Could you bring him here at about four o'clock?" asked Zimmermann.

"That would be perfect," said Martin. "Now, you should be aware that Carl is still just a young boy. He comes from a simple background, he speaks only in the local dialect, his mother is unable to read or write, and his father is just barely literate. Young Carl is shy, but still I think you will be impressed."

Martin and Carl arrived a few minutes before the appointed time the following Tuesday. Carl gazed at the impressive, three-story building that housed the college. It was an imposing building—the biggest Carl had ever seen. Büttner's one-room school was the only school that he knew, and he could see at a glance that this school had more than 20 rooms, all of them beautiful. There were paintings on the walls, the steps were made of marble, and Carl felt very small.

"Professor Zimmermann," began Martin Bartels, "I'd like to introduce Carl Gauss to you. He is the boy I was telling you about last week."

"Ah, yes. Carl is your name?" asked Professor Zimmermann.

"Yes, sir," Carl responded.

"I understand you are interested in mathematics," continued Professor Zimmermann.

"Yes, sir," Carl responded.

"Could you tell me the most interesting piece of mathematics that you have come across yet?" Professor Zimmermann asked.

"Oh, that is difficult. There are so many interesting pieces. I am curious about prime numbers. I believe there must be a pattern, but so far I haven't been able to find it. I know that the further out in the number system you go, the fewer primes you find, and I know there is no end to the set of primes. I can't believe that their occurrence is random, but I haven't been able to get beyond that. There is so much more that I need to learn," said Carl.

"If you could lead an ideal life, what would you do?" asked the professor.

"I don't know. I guess the first thing I would like to do would be to understand all of mathematics," said Carl.

"That, young man, is a tall order!" said the professor. "In a lifetime of study, I have only begun to understand some of mathematics."

"Only some of it?" asked Carl. "Well, at least I would like to see just some of the most beautiful patterns clearly. That would be exciting!"

"Yes, it would," said Zimmermann. "Tell me about some of the ideas you have been playing with."

"Well, *Herr* Professor, I have found some interesting things about the difference of two squares. I realized that I can apply it to multiplication problems that are too difficult to do in my head. If I want to multiply 22×28, that will simply be $25^2 - 3^2$ since it is simply $(25-3)(25+3)$. It is so much easier to subtract $625 - 9$ and get 616."

"That is true. Have you studied Latin?" asked the professor.

"Yes, Martin and I have been working through the grammar together," said Carl. "I like its logic, and in many ways translating Latin is like solving a good puzzle. In a way, it is mathematical."

"Yes, it is," said the professor. "*Herr* Bartels, I believe the Duke would like to meet your young friend. Thank you so much for introducing me to him. I'll let you know what I am able to arrange."

As the two boys left the professor's office, Carl was silent. He could hardly believe what he had just heard. Martin was not so surprised. He knew that Professor Zimmermann was an important man in Braunschweig and that he had some contact with the Duke. Martin had hoped that Zimmermann would be able to intercede for Carl with the Duke, but he hadn't been sure that Zimmermann would be willing to do that after only one interview.

"Martin," Carl asked quietly as they started back toward Carl's house, "what do you think the Duke will do?"

"I can't say for sure, but I hope he'll pay for you to go to the *Gymnasium*. That is where you can continue your education," said Martin. "It's time for you to move beyond the elementary school. Apparently Professor Zimmermann was impressed with you and will go to work on your behalf. We'll have to wait and see what happens."

"You know, when I was very little, the Duke's carriage hurtled past my mother and me and splashed muddy water on us. I was outraged! My mother said it was nothing—I couldn't believe it!—and she told me how important the Duke is. If he is that important, do you really think he will make time to do something for me?" asked Carl.

"The Duke wants the whole Duchy to thrive. The Duchy is made up of many individual people. The Duke can't know

Royal Patronage

all of them, but he certainly knows some of them, and when someone has genuine ability it is in the Duke's best interests to help that person get ahead. I hope you are that someone, Carl," said Martin.

"But certainly there are others far more deserving than me!" said Carl.

"Deserving is not the question," said Martin. "*Herr* Büttner and Professor Zimmermann and I believe you have the ability to do some very important work in mathematics."

"But I'm only 11 years old," said Carl. "How can anyone tell what I'm going to do?"

"Nobody can tell for sure," admitted Martin, "but we all think you might do something significant. Trust us."

Later in the week, Zimmermann had an interview with the Duke at the palace. The Duke depended on Zimmermann for help in a number of areas, but when the Duke had finished with his agenda, Zimmermann asked for permission to speak about a child in town. "Your Highness," he began, "I met a remarkable child from the Katherineum school this past Tuesday. He is 11 years old, and he has already mastered all of algebra and geometry. He seems to be more than ready to move on to the *Gymnasium*, but I believe his family does not have the money to pay the fees. His father is a simple stonemason. I thought you might be interested in meeting the child."

"You think the child is truly outstanding?" asked the Duke.

"I believe so, Your Highness. He is a child prodigy—I think a genius. He is possibly the brightest child that I have ever met," said Zimmermann.

"Well then, I would like very much to meet him. Can you bring him to me next Monday at three o'clock?"

"I would be pleased to do that, Your Highness," answered Zimmermann.

"Your Highness," began Professor Zimmermann, "this is the child I was telling you about: Carl Gauss. I believe he is a child with great ability, and I wonder if you would consider helping him get an education."

"Thank you, *Herr* Professor. Tell me, child, what would you like to do with your life?" asked the Duke.

"Well, sir, I am not entirely certain. I love mathematics. Of that I am sure. My greatest pleasure is learning and discovering ideas and exploring new ways to use the ideas I have discovered. I can imagine nothing better than spending my life learning and discovering fascinating new ideas."

"How would you do that?" asked the Duke.

"I love reading about new ways to think—are you familiar with Euclid's proof that the square root of two is irrational, sir?" asked Carl. "I mean, Your Highness."

"I can't say that I remember how it goes, although I must have encountered it once," said the Duke. "Could you please explain to me what it means for a number to be irrational?"

"I'll try, Your Highness," Carl began. "The square root of two is irrational because it is impossible to find a number or fraction that you could multiply times itself to get the number two. The square root of two would be that number, and you would think that with all the numbers, there would be a number that would work. If there were a number, we should be able to write it as the ratio of two whole numbers, like $\frac{3}{2}$ or $\frac{5}{4}$. But when we multiply $\frac{3}{2} \times \frac{3}{2}$, we get $2\frac{1}{4}$—and that's too big. If we try $\frac{5}{4} \times \frac{5}{4}$, that gives us $1\frac{9}{16}$, but that's too small. We could keep going and get closer and closer, but we would never find a number that we could multiply times itself to get exactly two.

"The way Euclid proved it, Your Highness, was to say that if there were such a number, he could write it as a ratio—a fraction—of two whole numbers that he called p and q. He said that if that fraction had already been reduced completely, there would be no number besides one that would go into both the numerator and the denominator evenly. So he wrote an equation: $\frac{p}{q} = \sqrt{2}$. Then he squared both sides of the equation, and he got $\frac{p^2}{q^2} = 2$, and then he multiplied both sides of the equation by q^2. That gave him $p^2 = 2q^2$, and that meant that p must be an even number. Does that make sense to you, Your Highness?"

"Yes. Keep going."

"If p is an even number, then there must be a number n such that $p = 2n$. So then he substituted $2n$ for p and he wrote $(2n)^2 = 2q^2$, and then he squared $2n$ and got the equation $4n^2 = 2q^2$, so dividing both sides of the equation by 2, he found that $q^2 = 2n^2$, but if that were true, then q was an even number, too. That would make both p and q multiples of 2, but at the beginning he had said that p and q did not have a common factor greater than one. That meant that the original premise was wrong, and as a result it was clearly impossible to find two numbers whose ratio was equal to $\sqrt{2}$. That is how Euclid demonstrated that the square root of two is an irrational number. Did that make sense, Your Highness?"

The Duke stared down at Carl in awe. "Yes, it did. Thank you. Is that the kind of thing you think about in your spare time?" he asked.

"Yes, sir. I guess that is what I like to do best," Carl admitted.

"How old are you, child?" asked the Duke.

"Eleven years old, Your Highness."

"Professor Zimmermann thinks that you have the ability to do some important work. Certainly the first step would have to be studying at the *Gymnasium*. Professor Zimmermann, would you please take care of arrangements for young Gauss to study there? I will underwrite the costs. It appears that he is willing to work hard and that he is intelligent enough to work effectively. Good luck, my boy! I look forward to seeing what you are able to do."

"Thank you, sir," said Carl. "Thank you very much!"

"Thank you, Your Highness," said the professor. "I appreciate your help. I think you will not regret this move."

When Carl got home that afternoon, his mother was eager to know what had happened. "Carl, did you meet the Duke?" his mother asked.

"Yes, *Mutter*," Carl answered. "He asked me what I want to do in my life."

"What did you tell him, Carl?" she asked.

"I told him how much I like mathematics, and I explained about irrational numbers," Carl said. "He seemed to like what I said. As a matter of fact, he told Professor Zimmermann to arrange for me to study at the *Gymnasium*. He will pay for it."

"Carl, that is so exciting!" his mother exclaimed.

Gymnasium: High School

The *Gymnasium*, the counterpart of an American college-preparatory high school, was located a kilometer from the Gauss' home. It was a two-story, stone building, and most people who were associated with it were connected in some way to Braunschweig's elite. Few children from the lower classes had the opportunity to study there, and those who did were at a disadvantage. There were no books in their homes, their parents had no understanding of what they were studying, and many of those parents were distrustful of the whole idea of an education. After all, they had been able to scrape out a meager existence with little or no education. However, each generation hopes that the next generation will enjoy greater success than they have, and, like them, Gauss' mother began to dream that her child would grow up to join the middle class.

Even though the Duke was paying for young Gauss' education, his parents still had to make sacrifices for him. He was not working in the home spinning flax, and he was not learning a trade in an apprenticeship to a local craftsman as most children of his age and class were doing. It looked as if the family would need to continue to feed and clothe him for some time even though the Duke was officially supporting him. It was an arrangement that Gauss' father did not particularly like, but he had agreed to it and he would follow through on his promise. His mother never questioned Gauss' potential, and Gauss always knew that she would do whatever she could to help him.

When Carl entered the *Gymnasium* in 1788, he began his serious study of Latin and Greek. Although he and Martin had already worked their way through Latin grammar, Carl needed to know far more than the grammar. He needed to be able to write it fluently since all scientific writing at the time was done in Latin. Because Greek was the language of the ancient mathematicians and philosophers, Carl needed to master it as well. He had a natural talent for languages, and he learned Latin, Greek, French, and English easily before he completed his studies at the *Gymnasium* in 1792. By that time, he and his classmates were writing notes in French or Latin, apparently without effort!

As a first-year student, Carl found himself in classes with other boys about his age. However, Carl already knew most of the mathematics that was being presented. Over time, they realized that he was far ahead of them. In his second year, some of the boys were talking one day before school. "Gauss, did you already learn trigonometry in elementary school?" Johann Ide asked in amazement.

"Yes, I had a tutor who helped me through a lot of interesting stuff. Martin Bartels and I worked through trigonometry and we were ready to start the calculus by the time I finished," Carl explained.

"Would you mind helping me a bit?" asked Johann. "I understand algebra and geometry, but trigonometry just doesn't make sense." The two boys spent several hours after school that day and several days in the weeks to come, with Carl explaining the difficult parts to Ide.

"Gauss, I don't know how I would have gotten through this mathematics without your help," said Johann.

"Oh, it's not so hard! You just have to think it through," said Carl. "Martin helped me a lot in the last couple of years, and I am glad to do the same for you."

"Yes, but you bring in some things that *Herr* Schmidt doesn't mention," said Johann. "Why couldn't Schmidt have told us that the distance formula is really nothing more than the Pythagorean Theorem?"

"I don't know. It just made sense to me to look at it that way," said Carl. "Mathematics is really fascinating. I spend a lot of time just thinking about it."

When Carl got home that evening, his father decided that it was time to talk with his son. "Carl, what do you do all day when you are at school? You already know enough to do plenty of jobs well."

"You don't understand, Father" said Carl. "What I need to do is investigate the important parts of mathematics. I need to study number theory. I need to understand and explain the fundamental theorem of algebra. I need to learn Latin well enough that I can write it easily. There is so much that I need to learn."

"Why is that?" his father asked. "It looks to me as if you aren't getting anywhere. You should be learning a trade so that you can support yourself."

"Father," Carl said, "my mathematics is terribly important. You must accept that."

"That's nonsense!" said his father. "I told *Herr* Büttner that I would let you do this, and I will not go back on my word, but I don't like it."

At about this time the Duke, acting on a suggestion from Zimmermann, arranged for Gauss to have a book of logarithmic tables to help him in his calculating. Logarithms were the only practical way to do multiplication of large numbers before the inventions of the hand-held calculator and the computer many years later. Because "logs" allowed Carl to do large multiplication problems by adding the exponents on the number e instead of doing the tedious multiplication, these tables expedited much of the mind-numbing calculating Carl compelled himself to do. Carl's tables of "logs" were a constant companion to him, and, by the time he was 20 years old, he knew most of them by heart. Later in life Carl claimed to find poetry in the log tables!

The Duke also instructed Zimmermann to provide Carl with a set of drawing instruments including several compasses and a fine brass straight-edge. If Carl was to study Euclid seriously, he needed to be able to make accurate drawings. Now that Carl had the equipment necessary to master mathematics, he plowed ahead, going far beyond any content that was expected of him in the *Gymnasium*.

A real challenge for Carl at the *Gymnasium* was learning High German. In his home and neighborhood, everyone spoke the Braunschweig dialect. When he read books, they were in High

Gymnasium: High School

German, but he had never heard anyone except his teachers, Professor Zimmermann, and the Duke talk that way. There were striking differences between High German and his dialect, but, because it was basically the same language, it was sometimes hard to recognize the differences and see his mistakes. Sometimes it was only a question of pronunciation, but other times the words and the grammar were completely different. That meant he had to be alert all the time. *Herr* Büttner had touched on a few details of High German, but he presented them only as isolated facts. Carl would not be taken seriously outside of Braunschweig if he slipped into dialect as he spoke or wrote.

"*Herr* Gauss," said the German teacher as he returned an essay that Carl had written, "you have written about the writer Gotthold Ephraim Lessing's drama *Nathan The Wise*. You describe Nathan as a smarty-pants. I don't like your choice of words. Lessing is a famous author who lives in Wolfenbüttel in our Duchy of Braunschweig. When you refer to Nathan's wisdom, you should not write in dialect. In formal writing, you must use a literate vocabulary. You could describe him as intelligent or brilliant."

"Yes, sir," said Carl. "I'll try to be more alert as I write."

Carl spent much of his time reading mathematics. Euler, the great Swiss mathematician, fascinated him. He wondered how Euler could have written so much on so many subjects in mathematics! Even when Euler was blind during the last 17 years of his life, he still wrote quantities of original mathematics.

Carl also studied Archimedes and Lagrange.

However, his hero was Isaac Newton, the Englishman who was one of the two discoverers of the calculus. Newton's great work, *The Principia*, is often seen as the finest mathematical exposition of physics ever written. Gauss studied all that Newton had written, and he mastered it all. Since Gauss' own time, Gauss himself has often been compared with Newton. This would have flattered Gauss immensely. Newton attributed his brilliant discoveries to the fact that he was able to see further by standing on the shoulders of giants—the mathematicians who came before him. Gauss also stood on the shoulders of giants, including Newton's.

Gauss spent hour after hour, day after day, trying to find the pattern of primes and their distribution through the number system. Johann Lambert in Switzerland had compiled the table of prime numbers that Gauss used, but Gauss checked enough of Lambert's calculations to convince himself that Lambert was not totally reliable. He was outraged by Lambert's errors—he described Lambert's tables as "bristling with inaccuracies!" In fact, Gauss' corrected version of Lambert's table was excellent—very close to the figures found with the help of computers by the end of the twentieth century.

"Mr. Newton," Gauss told the mentor in his mind, "I know the primes get rarer and rarer the bigger the numbers get. I wonder what would happen if I explore how many primes there are below any given number. I could call it 'π of n' letting π [the Greek letter p] stand for the number of primes and n for how many numbers I'm talking about. Below 100 (when $n = 100$), there are 25 primes, so $\pi(100) = 25$. Below 500 (when $n = 500$) there are 95 primes, so $\pi(500) = 95$. Using that notation,

$\pi(1000) = 168$. There are always more primes the further I go, but the number $\pi(n)$ is growing more and more slowly—in general we add fewer new primes for every new set of 100 numbers.

"No, wait a minute! Instead of looking at the total, why don't I look at the number of new primes I find within each range? While there are 25 primes between one and 100, there are 21 primes between 100 and 200, and 16 primes between 200 and 300. If I skip up to higher numbers, there are 12 primes in the 100 numbers between 5,000 and 5,100 and only 9 primes in the numbers between 20,000 and 20,100. Yes, the number of new primes I'm finding is decreasing, but it's decreasing very slowly. I wonder if I could graph those figures. What would it look like?

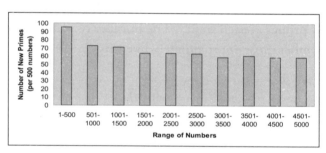

"Do you know what, Mr. Newton? Doesn't my graph remind you of the reciprocal of the natural logarithms:

$$\frac{1}{\ln(n)}?$$

"I think that's what it is. Wow! I don't know how I can prove it, but I think that explains it. If I'm right, and I think I

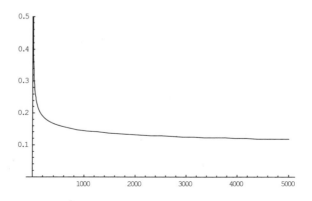

The graph of the reciprocal of the natural logarithms

am, then we could find out how many primes there are below a certain number (that's π(n)) by simply adding all those numbers (the new primes at each step)." The number of primes below any given number grow approximately like this:

$$\pi(n) \sim \frac{n}{\ln(n)}.$$

This is now known as the Prime Number Theorem, and Gauss apparently discovered it in 1792 when he was just 15 years old. Although the published tables of primes had been available to serious mathematicians for many years, only Gauss was able to discover the pattern as he played with those tables. Mathematics had to wait another 100 years for Hadamard and de la Vallée Poussin to finally prove it.

Gymnasium: High School

One afternoon after school, Gauss and his friend Ide were walking home. Before Ide turned toward his house, Gauss commented, "Ide, I've been trying to measure the distance from my house to the *Gymnasium*, and I've been counting my steps every day for two weeks. Would you believe I have gotten a different number every day? Since the paving stones are uneven, sometimes they make me take a shorter or longer step, but I've been trying to compensate for that. Still, I keep getting a different number. It's mystifying."

"I can't say I ever worried about it," confessed Ide. "Why do you need to know?"

"Well, I don't need to know, but wouldn't you think I could figure it out? Every day I think that I am keeping a constant pace, but apparently I am not succeeding. The number of steps between my house and the *Gymnasium* is not the point. I'm interested in measuring and consistency. I think I ought to be able to get the same number of steps every time." Reaching into his satchel, Gauss took out a piece of paper. "I've also been measuring the length of this line segment on this paper using the new metric system. Even though I try my very best to be precise, I keep getting different answers," said Carl.

Stopping to set his satchel down, Ide offered: "Here, let me try it. Okay this line segment is 13.5 cm long. What measurement did you get?"

"Well, the first time I tried, I got 13.6 cm, the second time I got 13.4 cm, and the third time I got 13.4 cm again, but then the next time I got 13.5," said Gauss.

"Oh! I didn't realize you were being that precise. Let me try again," said Ide. "Um, this time I get 13.4 cm. Let me do it again: 13.5. Maybe you have a point. I'm being as precise as I can, and I'm not getting the same measurement twice. That's amazing. Why do you suppose that happens?"

"I don't know. To err is human, and all that? Maybe it's because we have to align both ends of the ruler and there is a possibility of missing by a little bit at each end. But don't you think that two smart boys like us ought to be able to find the exact length of this segment, Ide?" asked Carl.

"I would think so. I wonder if everyone has this kind of trouble with measuring," wondered Ide.

"It's possible that this is a new problem because we have more precise measuring scales. Last night as I was measuring the segment, I discovered some interesting patterns. All my measurements hovered around 13.5. Some were a little more, and some were a little less. Your measurements do the same thing. I also noticed that I got about as many numbers less than 13.5 as I got bigger than 13.5. Most of my measurements were very close to 13.5, but a few of them were further away, and the further the measurement got from 13.5, the fewer numbers I got. Finally the measurements petered out to nothing. When I plotted my measures on a graph, they almost seemed to be arranged in a bell shape, with most of the measurements close to 13.5, and then they trailed off to zero when I got to 13.7 or 13.3. When I just averaged them to try to get the big picture, I got a number that looked pretty good, but I think that is not the best way to give the data.

Carl continued, "I wonder what would happen if I were to choose some value that looks pretty close to the true value and then I'd find the difference between that 'true value' and each measurement," said Carl. "You know what? If I squared all those differences, the ones that are further away would count more and that would also make all the numbers positive. I think that might be better. If I added up all those squares, I would look at the total. Next I could try another guess at the 'true value' and see how the sum of the squares of the differences compared to the sum of the squares using the other 'true values.' I would choose the 'true value' that gave me the smallest sum. I've got to find a more efficient way of doing this, but I think the principle is right. I think I'll call it my 'method of least squares.' I suppose lots of people have used this for years. Oh, well. Now I can use it too."

"You know, Gauss," said Ide, "I don't remember seeing anything about this in any of my mathematics books. Why don't you ask *Herr* Schmidt about it? He might find it interesting."

"Oh, no. I wouldn't bother him with this. I'm sure it's something he's always used."

"I wouldn't be so sure, Gauss," said Ide.

"Oh well," said Carl. "I think I'll wait to discuss this with anyone else for awhile yet. I think I need to do some more work on it first."

In fact, Gauss was the first person to use the method of least squares. Legendre discovered and published it in 1805, at least ten years later. In research, the first person to publish a discovery gets the credit, so Gauss never asked for recognition. He knew the rules.

Independent Study of Mathematics

When Carl was 15, he began studying at the Collegium Carolinum, where Professor Zimmermann taught. Soon after Gauss entered the college, Professor Zimmermann determined that there was no need for Gauss to take any mathematics courses there since he already knew all the mathematics that was in the curriculum. As a result, Gauss concentrated in his classes once more on classical and modern languages. However, he did not ignore mathematics during these years. In fact, he continued to explore mathematics intensively wherever his curiosity led him. Like so many mathematicians before him, Gauss still wanted to prove Euclid's parallel postulate (through any point you can draw only one line parallel to a given line), and like them he couldn't do it.

"I wonder," Carl mused, "would it be possible to approach the parallel postulate in a different way? Hmm. What if I come at it from the opposite direction? What if it were possible to draw more than one line parallel to a given line through a given point. What would that look like? How would I construct it?

"Aha! What if the surface is not necessarily a flat plane? What if it is a curved surface, like a saddle for a horse or the surface of a ball. Let's look at the earth. The lines of latitude are parallel, but the lines of longitude are not. At the North Pole, all the longitude lines come together, even though they seem parallel from our limited perspective. Well, I believe I've got something. And what do I mean by parallel? That they just look parallel? I don't think so. Parallel lines are two lines that will never meet no matter how far they are extended. That definition will work in plane geometry—lines that look parallel will never meet—but it will also work on my warped surfaces.

"So that means, that I can construct a totally new geometry on nonflat surfaces, where there are points and lines and circles and there are parallels, but now some of the parallels look a little different. Yes! We can have a line and then a point some distance from the line. On the plane, we can draw only one line parallel to the first line through that point that is off the line. But I could also distort the plane a bit, causing it to bulge and stretch. This time, I could draw a different line through the point and still parallel to the original line. My new line still passes through that same point, but it is not the same line that I drew in plane geometry. I could warp it again and find another line that is parallel to the original line. This doesn't conflict with Euclid's first four postulates, and it definitely lets me draw more than one line to meet the criteria.

"You know what, though, Mr. Newton? I'll bet this is not what Zimmermann expects me to explore. No, I don't think mathematics is ready for this yet, but I'm sure I'm right. I'll wait until the mathematical world is ready for this. However, I don't need to waste any more time on proving the parallel postulate using the first four postulates of Euclid. Obviously it can't be proven."

One day Professor Zimmermann saw Gauss coming down the hall and asked him to see him in his office. "*Herr* Gauss, I hope your studies are going well," the Professor began.

"Yes, *Herr* Professor," Gauss answered. "I believe I have finally gotten to the point that I can write Latin pretty well, and I have actually become a real fan of Greek language and literature. I hadn't realized what I was missing before I learned Greek."

"From what I hear, you have done very well in your language classes. That pleases me."

"I hope I do well enough, *Herr* Professor. I want to be sure to live up to your expectations and the Duke's," said Carl.

"Oh yes! There is no need to worry about that," said the Professor. "Your work has been outstanding. However, the Duke sent me a message today, asking me to have you come talk with him at the palace tomorrow afternoon."

"Really? Do you know what he wants to see me about?" asked Carl uneasily.

"My suspicion is that he feels that you are ready to move on to the university. I certainly believe you are ready. I particularly want you to study in a place that can nurture your mathematics ability better," the Professor admitted.

"Oh you don't need to worry about that, *Herr* Professor. You have been a wonderful inspiration and support for me during my years at the college," said Carl.

"That may be, but I believe you are ready to move on. Go see what the Duke has to say."

"All right," Carl answered. "Thank you, *Herr* Professor, for your help in this and everything else."

The next day, Gauss walked to the palace. When he arrived, he asked the servant to tell the Duke that Carl Gauss was there for his appointment. A few minutes later, he was ushered into the Duke's library where he found the Duke sitting behind an elegant desk made of ornately carved oak. He sat in a large chair with intricate carving on the back and arms.

"Tell me, *Herr* Gauss," the Duke began, "what you would like to do next. You have done well at the college, but I think it is time for you to go to the university. Professor Zimmermann tells me that at the college you have not been able to study mathematics, your real passion, because you had already moved beyond all that is offered at the college even before you left the *Gymnasium*."

"But Your Majesty," Gauss answered, "please don't think I have been ignoring mathematics! I have done a great deal of mathematics while I have been studying at the College, and Professor Zimmermann is always generous with his time whenever I need to discuss something that I am working on. I can assure you I have not neglected my study of mathematics."

"No, I didn't mean to imply that you had," said the Duke, "but I think it is time for you to move into a larger academic environment. I would like you to go to our university at Helmstedt. That is the university for our duchy, and I flatter myself that it is an excellent university. I believe you would thrive there."

Independent Study of Mathematics

"Actually, Your Highness," Gauss said, "I was hoping that when the time came you might be willing to consider sending me to the university at Göttingen instead. I know it's farther away and not in the Duchy of Braunschweig. The problem is that I have already read all the mathematical literature here at the college, and, according to Professor Zimmermann, the library at Helmstedt has roughly the same collection. Göttingen, however, is supposed to have a vast mathematical collection, and I believe it would serve me far better. I need to read Euler and Newton and Descartes in the original, to say nothing of the oldest extant versions of Euclid and Diophantus, and many of those resources are simply not available at Helmstedt or anywhere else in Braunschweig."

The Duke replied, "*Herr* Gauss, that makes excellent sense to me. Yes, you may go to the university at Göttingen. You'll need a bit more money there, but I will arrange for that as well as for food and rent. You have done extremely well here at the college, and I know that you will continue to grow in Göttingen. I expect you to become a great scholar."

"Thank you, Your Highness! Thank you so much! I will do my very best not to disappoint you, sir."

"Excellent. Goodbye for now, *Herr* Gauss," said the Duke.

"Goodbye, Your Highness, and thank you."

The Regular 17-gon

When Gauss moved to the town of Göttingen, he had to learn his way around the thriving university town. Although the town was smaller than Braunschweig, it was Gauss' first time away from his hometown, and he found it overwhelming at first. He was able to rent a room above a store on a small street near the university, and he quickly developed a mental map of the town and university. It was an adventure in every way, and he was determined to make the most of it.

After the first week of classes, Professor Seyffer invited a few of the new students for supper in his home one evening. Professor Seyffer's home was a two-story row-house two blocks from the university. It was a pretty house with comfortable but tasteful furniture. Gauss had always assumed that his godfather, Herr Ritter's, home was as beautiful as any private home could be, but Professor Seyffer's home was far more elegant. "So tell me, *Herr* Gauss," began Professor Seyffer, "how you are finding our university at Göttingen."

"This is wonderful place to study, *Herr* Professor," answered Carl. "I sometimes feel a bit daunted by the size of the university, but I'm getting used to that. The town of Göttingen is small but big enough that I can always find what I need."

"How about you, *Herr* Bolyai?" asked the professor. "You come from Hungary, I believe. Is this your first time studying at a German-speaking university?"

"No, *Herr* Professor," said Wolfgang Bolyai. "I studied for two semesters at Jena, but Göttingen is different. The whole atmosphere is so exciting. Here I can learn and grow. I like it very much."

"*Herr* Bolyai," asked the professor, "what are you interested in studying here at Göttingen?"

"Mathematics is my favorite," said Bolyai. "I would like to make a career in mathematics. I would like to do some original research."

"And you, *Herr* Gauss?" asked Seyffer.

"I haven't decided completely," Carl admitted. "I am drawn to mathematics, but I am also fascinated by the study of languages. For the moment, I would like to keep both options open. I am only eighteen years old, and I hope that I still have the time to explore both fields before I have to commit to one or the other. I realize I will have to make a choice before long, but I would like to put it off for a bit."

A few days later, Bolyai called to Carl whom he saw walking along the old town wall, and they began to talk: "*Herr* Gauss, I have been working on the parallel postulate of Euclid, and I was wondering if you have done any work on it."

"Yes, I did some work on it several months ago. Have you made any progress on it?" asked Carl.

"Well, I have been playing around with an argument involving the perpendiculars, and I would like to think that I am getting somewhere with it," Bolyai said.

Carl was delighted: "You are a genius, my friend." Carl saw no need at this point to explain his own discovery about the parallel postulate. After all, he had not spent as much time on it as he would like to, and he didn't know Bolyai well. Also, it was possible that Bolyai would probe the problem from a completely different angle, and Carl would love to have a genuine dialogue about geometry. If he presented his discovery first, it might limit the possibilities later. Besides that, Gauss had determined to wait to act on his discovery until he felt that others were ready to hear his revolutionary thoughts. Carl could tell from Bolyai's comment that he was using a more traditional approach.

"*Herr* Gauss," Bolyai continued, "you are interested in mathematics, aren't you?"

"Oh, yes," answered Carl. "It is a beautiful subject."

"Have you studied at another university before coming here to Göttingen?" asked Bolyai. "I get the impression that your background in mathematics is more complete than mine."

"Well, I studied at the Collegium Carolinum in Braunschweig for almost four years," admitted Carl.

"What mathematics did you study there?"

"Actually, I didn't study mathematics there. My grounding in mathematics from my *Gymnasium* days made the mathematics courses at the college redundant for me," said Carl. "What I did for the last four years was to explore mathematics on my own, and then I consulted with Professor Zimmermann at the college when I needed help refining a point."

"Does this mean that you taught yourself mathematics?" asked Bolyai.

"I don't know if that's the way to say it," said Carl. "I read widely in mathematics in the college library, and I investigated the questions that arose from my reading. I am particularly impressed with Newton, and I have spent many hours studying the *Principia*. I am trying to find Emilie du Châtelet's translation of Newton's *Principia* into French, but the library at the College didn't have it, and I haven't been able to find it at the university library here either. I understand that du Châtelet found some interesting corollaries to Newton's work and that she built a sort of bridge between Newton's calculus and Leibniz's. It looks as if it has been a mistake to view her only as an appendage of Voltaire. Her understanding of mathematics was far more penetrating than Voltaire's."

"I am not familiar with her name. Was she French?"

"Yes, she did her translation of Newton about 50 years ago."

"If you are able to find her work, I would love to look at it," said Bolyai. "I suppose you are able to read French."

"Yes, she wrote in French, and that won't be a problem for me" said Gauss, "but if I could find it in English, that would be okay too. For that matter, if it were available only in Russian, I would gladly learn Russian so that I could read it."

"Are you attending Kästner's lectures in mathematics?" asked Bolyai.

The Regular 17-gon

"Yes, although he doesn't give me very much to attend to," said Carl.

"You are not impressed with him?" asked Bolyai.

"No," said Carl. "Are you?"

"I have to admit that I am a bit disappointed," said Bolyai.

"I am more than a bit disappointed. I came here to study mathematics and languages, and I am delighted with the mathematics collection in the university library (even though they don't seem to have du Châtelet), but I am not impressed with Kästner."

At that time, studying at a university was a rather casual affair. There were no requirements and no deadlines, one could attend lectures if one chose to, and ideally one learned. Because Gauss was always thinking about difficult concepts and always discovering new approaches and new ways of looking at his own ideas, this learning style fit him beautifully. When he was exploring an idea, the university library often had books that could show him what mathematicians before him had done on that subject, and if he didn't find it he could be fairly sure that the topic had escaped the attention of others. He found the university a congenial place to explore mathematics and many other fields, and he was intellectually alive.

A few months later, Carl and Bolyai sat together for Kästner's lecture. Afterward they strolled through the park. Carl had something on his mind: "Bolyai, do you remember when I told you and Seyffer that I couldn't decide whether I should concentrate on languages or mathematics?"

"Yes I do, Gauss, but I have to admit it surprised me. You seem so utterly fascinated with mathematics!" admitted Bolyai.

"Well, it's true, but I also love languages," said Gauss. "You know, people express their deepest thoughts and create remarkable new ideas only through the gift of language. I am amazed at how differently that process works in different languages. However, I believe I have made up my mind now: you were right about my commitment to mathematics. You will never guess what I have been able to do! Look at this drawing. It's a regular 17-gon, and I made it using only a compass and straight-edge. I had to do a lot of figuring, but it is safe to say that I can construct a regular 17-gon—in fact a polygon with any number n of sides where n is a Fermat prime number."

"Fermat primes...," interrupted Bolyai. "Fermat primes are one less than a power of two, aren't they? Aren't they numbers like $2^2 - 1 = 3$ or $2^3 - 1 = 7$?"

"No, Bolyai, those are the Mersenne primes," said Gauss. "A Fermat prime is totally different, although the number three is a Mersenne prime as well. For a Fermat prime, you raise two to a power of two and then you add one to that: $2^{2^0}+1 = 2^1+1 = 3$, or $2^{2^1} + 1 = 2^2 + 1 = 5$, or $2^{2^2} + 1 = 2^4 + 1 = 16 + 1 = 17$, or $2^{2^3} + 1 = 2^8 + 1 = 256 + 1 = 257$. There are very few Fermat primes. In fact, there are probably only four of them in all."

"And there are a lot more Mersenne primes, aren't there?" said Bolyai.

"Yes, people will probably continue to find new Mersenne primes forever," said Gauss. "At any rate, with Fermat primes, using a compass and straight-edge I can construct a regular poly-

The Regular 17-gon

gon with 3 sides, in other words an equilateral triangle. I can construct a regular polygon with 5 sides, in other words a regular pentagon. I can construct a regular polygon with 17 sides as I just did here. I could also construct a regular polygon with 257 sides, although fitting it on a piece of ordinary paper drawn to any scale that would be visible would be difficult, to say the least. I could also make a regular n-gon using any multiple of Fermat primes if I wanted to—like a decagon or a 15-gon or a 34-gon. The only limitation is that the Fermat number n that I start with must be prime."

"Now, wait a minute, Gauss! You're smart, I've figured that much out. But are you that much smarter than Euclid or Euler? Remember, they said a 17-gon couldn't be constructed with a compass and straight-edge."

"No, Bolyai, they didn't say it couldn't be done. They just didn't figure out how to do it themselves. I'll bet they never even considered making a 17-gon," corrected Gauss.

"I expect that's true," Bolyai admitted.

"Well," Gauss replied, "maybe they didn't expect anyone to do it, but I know it can be done because I have done it. I used a good bit of trigonometry to figure it, but it's all here."

"May I take it back to my room and see if I can follow it?"

"That's what I was hoping you would do. Oh, by the way, I'm meeting Wilhelm Eschenburg and Johann Ide, my friends from the college, this evening at the *Kneipe* [tavern] across from the library. We thought we'd meet at about 8 o'clock. Won't you join us?"

"That sounds great!" said Bolyai. "I hear they are both fine students of mathematics."

"Yes, and Ide and I share something besides mathematics: the Duke of Braunschweig is supporting both of us in our education. We both come from poor families, but the Duke is determined that we should get a good education. The Duke has certainly done a great deal for both of us. Eschenburg is different. His father is a professor at the college. He doesn't need any help from the Duke."

"It would be fun to get to know them a bit. Shall I bring the 17-gon with me?" Bolyai asked.

"No, I don't think so. I'd like you to study it and tell me what you think. Then I'm going to send it to Zimmermann, my mathematics professor at the college. We'll let Eschenburg and Ide see it for the first time in print."

"That's fine!" Bolyai answered. "This afternoon I think I'll try to get some of the reading done for Kästner's class."

"The reading is not bad," said Carl, "although I had hoped that he would cover some material that I haven't already worked on. That doesn't appear likely now. I had already done this week's reading long before I finished the *Gymnasium*. Kästner really is the leading mathematician among poets and the leading poet among mathematicians. He hasn't told me anything that I didn't figure out years ago! I have finally concluded that he may be a fairly witty man when he is talking outside of the field of mathematics. But all that wit leaves him when he does mathematics. Isn't that odd?"

"Yes, it is," agreed Bolyai.

"Are you attending Heine's lectures in philology?" Carl asked.

The Regular 17-gon

Gauss' drawing of Kästner lecturing. Note the deliberate arithmetic error

"No," said Bolyai. "Is he good?"

"Oh yes!" said Carl. "His study of philology is a linguist's dream. His interpretations of the classics set the standard for the study of the Greek poets. When I was at the library yesterday, I checked out Aeschylus' *The Oresteia*, and Heine is right. I like his interpretation. He is a scholar I can respect. If he is presenting a lecture, I will be in the front row so I don't miss a thing."

"Interesting," said Bolyai. "I think I'll try to start going to his lectures."

"By the way," Carl said, "I have heard many good things about Professor Pfaff, the mathematician at Helmstedt. I think I may go over there and study with him before I am done. I came here to Göttingen for the library, but certainly not for Kästner. I believe I have already gotten all that I'll ever get out of him."

The entrance to the *Kneipe* was at the corner of Pauliner Street and Papendiek Street. It was popular with university students and faculty alike. Ide stepped down the three steps to enter the low, dark room, partially illuminated by candles. Everywhere around him, there were heavy wooden tables with benches. Many were already full, but he found a free table, shed his warm coat, ordered a mug of cider, and took out the problem he was working on. Soon Gauss joined him.

"Ide, how's it going? Is that the silly problem Kästner tossed out today?"

"Yes, Gauss. This is it. I can't make sense of it."

"With good reason," Gauss rejoined. "The problem he gave us can't be solved using the set of real numbers. You see this number? It should be negative—not positive. Now I think you'll be able to get it. You've met my friend Bolyai, haven't you? I asked him to join us tonight. He is a fine fellow and also a fine mathematician."

"Yes, you introduced me to him a couple of weeks ago. He's very shy, isn't he?"

"He doesn't need to say much," said Carl, "but what he says is always important. Here he is now. Bolyai, come join us! Did you discover the glitch in Kästner's problem?"

"Yes, it was a pity. He should be more careful than that. At first I was stymied by it, but, when I saw his error, the problem practically solved itself. I guess your opinion of him hasn't changed, has it Gauss?"

"Oh, yes it has. It has reached a new low. I will probably skip his lectures from now on. I have important work to do, and his lectures are nothing but an interruption. Here's Eschenburg! Eschenburg, come join us!" he called to his friend.

"Thanks," Eschenburg said in greeting. "What do you have in that mug, Ide? Cider? What a good idea! *Herr Ober* [waiter], please bring me a mug of cider. Gauss, don't you want one too? Bolyai? Okay, then. Three mugs of cider, please. Listen, guys, I've been doing a little work on prime numbers, particularly Fermat primes. Do you think there are any Fermat primes beyond $2^{2^4} + 1$? That was probably a silly question. I don't know how anyone would be able to prove it anyway since a new one would have to be so huge."

"Nobody has been able to find any Fermat primes with an exponent greater than four yet," said Carl, "but show us what you've been doing, Eschenburg."

"Okay. I started with $2^{2^0} + 1 = 2^1 + 1 = 3$, $2^{2^1} + 1 = 2^2 + 1 = 4 + 1 = 5$, $2^{2^2} + 1 = 2^4 + 1 = 16 + 1 = 17$, and $2^{2^3} + 1 = 2^8 + 1 = 256 + 1 = 257$, and I know all four of those numbers are prime. It's generally agreed that $2^{2^4} + 1 = 2^{16} + 1 = 65,536 + 1 = 65,537$ is also prime although I certainly couldn't prove it. From there, Euler says he doesn't think there are any more that are prime although he couldn't prove it. I don't see how anyone can work on factoring numbers that big. Gauss, have you spent much time on primes and their patterns?"

"Well, actually I have. I figured out a few of years ago the basic distribution of primes through the set of whole numbers, but I haven't been able to prove it yet. As far as factoring big numbers goes, the bigger the number is, the harder it is to be sure it doesn't have any factors other than itself and one. Do you know what $2^{2^5} + 1$ is, Eschenburg?"

"Oh, *danke* [thanks], *Herr Ober*. Mmm. This cider is good!" said Eschenberg as he set his earthenware mug of cider down on the table. "Yes, I've got it right here. $2^{2^5} + 1 = 4,294,967,297$. Shall we give you five minutes to find its factors, Gauss?" Eschenburg asked playfully.

"It would have taken me more than five minutes if Euler hadn't found that it is the product of $641 \times 6,700,417$. The sixth power would take me quite a bit longer since I don't already know that one."

"Gauss, how do you remember facts like that? I find myself starting over every time because I just don't remember my earlier calculations."

"You know what I've noticed?" Bolyai interjected. "Gauss seems to think in mathematics harder and deeper and longer than the rest of us. I don't know anything about your mathematical accomplishments, guys, but I know that I for one can't focus on one problem as long and hard as Gauss does."

"I don't think I could do it even if I could focus on it that long," admitted Eschenburg. "Gauss, you do remember everything exactly as it appeared on the page after just one reading, don't you?"

"I suppose in a way I do, but I think Bolyai is right. I didn't just glance at those numbers. That factorization made a big impression on me when I found it, and I have to confess that I multiplied 641 × 6,700,417 out to be sure that it was right."

When Gauss and Bolyai met the following day, Bolyai had some questions. He had stayed up late the night before, trying to follow his friend's reasoning on the 17-gon.

"Gauss, I'm not saying you're wrong, but you lost me a few times. Show me how you got from this line to this," pointing to one section of the argument. Gauss was able to explain it all to Bolyai, who was impressed. Before this week, Bolyai had known how to make simple constructions like a regular hexagon or pentagon using a compass and straight-edge, but he had been happy to accept the fact that Euclid couldn't make any odd-ball regular polygons like a 17-gon, and neither could he.

"You know, Bolyai, I think when I die I would like to have a regular 17-gon carved on my gravestone."

"Like Archimedes?"

"That's what I was thinking." Archimedes, the Greek mathematician of antiquity, had discovered that a sphere that fits perfectly inside a cylinder always has exactly $\frac{2}{3}$ the volume of the cylinder. He had asked to have a drawing of this great discovery carved on his gravestone. Two hundred years after Archimedes' death, the Roman statesman Cicero wrote about finding Archimedes' gravestone in Syracuse under a mass of brambles, with no

Gauss family grave

inscription, but with the important drawing still intact. Unfortunately Gauss did not get his 17-gon on his gravestone because the stone carver refused to carve it, saying that people would think he had been trying unsuccessfully to carve a circle. He had too much pride in his work for that!

Mathematical Journal

"Mr. Newton, I have decided that I need to write down my mathematical thoughts as they come to me, so I have begun a mathematical journal. Sometimes as I'm walking around town or when I'm lying in bed when I first wake up in the morning, I am bombarded by several different ideas, all at the same time. There is no way I can go into all of them at once, and I worry about forgetting them. In fact, I'm afraid I have already lost some! I'll bet this happened to you too—I mean the ideas just coming to you out of nowhere, one after another. Before I got up this morning, three or four different thoughts came to me. When I wrote them down, I was able to start working on one of them without worrying about the others because I can always find them again in my journal. I think it's going to make a big difference.

Gauss called his now famous journal his *Notizenjournal* [Diary of Notes]. It is a small, bound notebook approximately $5'' \times 8''$. In fact, no one (not Bolyai and probably not even Gauss' wives or his children) knew of the existence of this journal until 40 years after his death, so it was truly a personal diary. The journal covers the years 1796 to 1814, although most of the 121

Notizenjournal: Gauss' personal journal of his mathematical discoveries, 1796–1814

entries date from the five years 1796–1801 when he was studying at Göttingen and during his sojourns in Braunschweig and Helmstedt. At this time, paper was expensive, so Gauss used it sparingly. His entries were concise, almost to the point of being incomprehensible, even to skillful mathematicians. On the 19 cramped pages, Gauss wrote cryptic notes in Latin, describing his mathematical discoveries.

Occasionally he made a comment similar to Archimedes' "Eureka!" about a discovery, as he did when he wrote $\Delta + \Delta + \Delta = N$. This is part of number theory, the field that Gauss called the Queen of Mathematics. A triangular number Δ is a number that you can get by adding up a string of consecutive whole numbers beginning with one. The number 6 is triangular because you can get it by adding 1 + 2 + 3, and 15 is triangular because you can get it by adding 1 + 2 + 3 + 4 + 5. The number 5,050, the sum of the first 100 counting numbers that he discovered when he was ten years old, is also a triangular number. The statement $\Delta + \Delta + \Delta = N$ says that every whole number can be written as the sum of no more than three triangular numbers. $29 = 28 + 1$ and $96 = 78 + 15 + 3$. To Gauss, that was a beautiful discovery.

Sometimes there is no indication in the journal of what great discovery it was or which "dragon" he had succeeded in slaying. As scholars have studied the journal, they have used it to justify Gauss' claims that he had devised a certain mathematical proof or operation long before others. It is the journal that shows that Gauss actually was the first person to construct a non-Euclidean geometry as well as the method of least squares although he said nothing to anyone about either discovery for many years. Nevertheless, when he learned of someone's else's discovery and he commented that he had known that for years, the journal proves that Gauss was speaking the truth.

The first entry in his journal begins in Latin: *"Principia quibus innititur sectio circuli, ac divisibilitas eiusdem geometrica in septemdecim partes etc.,"* which translates to "The principle depends on the geometric division of the circle into seventeen parts, etc." Gauss was justifiably proud of his accomplishment, and he wanted his journal to know it.

Now that Gauss saw himself growing as a mathematician, he adopted a motto for himself: *Pauca sed matura* [Few but Mature]. He devised a seal as well, showing an apple tree with just seven large, perfect apples. The motto and seal symbolized his determination to publish his mathematical discoveries only when they were perfect. It meant that each discovery should stand on its own, and there should be no indication of the long ordeal that might have been necessary to reach it. He wanted others to see only the perfect fruit of his labors. He believed that the concepts he presented to the public should be so complete that *nihil amplius desiderari posit* [nothing more could be desired]. His hero Newton also published very few of his results, but Newton's motive was different. Newton guarded his discoveries jealously, not wanting other people to get their greedy hands on them, while Gauss preferred to wait until the work was perfect. When Gauss was satisfied, he was eager to publish it unless he feared that the results would disturb the community of mathematicians. Since he didn't like controversy, that aversion sometimes deterred him from publishing until he thought the world was ready for it.

He drew an analogy between mathematics and the construction of a magnificent cathedral. The cathedral requires complex scaffolding to support the various parts during construction, but, when the building is completed, the scaffolding is removed and the building stands on its own. The journal gives only an occasional glimpse of the scaffolding necessary to produce Gauss' mathematics. He didn't expect anyone but himself to read the journal, and the barest essentials were all that he required. Mathematics was like poetry to him. Unfortunately, what was essential to him was much sparser than what has been essential to most mathematicians since.

In 1796 Gauss wrote an announcement of his discovery of the method for drawing a 17-gon using only a straight-edge and compass. He sent it to Zimmermann in Braunschweig with a request that Zimmermann could perhaps arrange to have it published. Zimmermann was impressed and immediately submitted it to a local journal for publication. That publication established Gauss as the discoverer of this technique, and it alone was enough to guarantee that Gauss would be considered one of the world's greatest mathematicians.

The 17-gon was only the beginning of an amazing career in mathematics. Gauss was only 18 years old when he discovered it. He lived and breathed mathematics for another 60 years.

Number Theory

"Bolyai," Gauss commented to his friend when they met one day outside the library, "I've been thinking about the decimal expansions of some rational numbers. We know that $\frac{1}{2}$ = .5, which is a terminating decimal. You and I know that $\frac{1}{3}$ = .33333333.... which is a repeating decimal because the digit three will keep repeating forever. We know that the fraction $\frac{1}{7}$ has six digits (142857) that keep repeating forever and ever:

0.142857142857142... or $0.\overline{142857}$.

I've worked out the decimal equivalents of the reciprocals of all the whole numbers 1 – 1000. I have to admit that it has taken me awhile because I had to carry some of them out to hundreds of decimal places. I can now prove that the number of digits that repeat (the period of the repetend) will always be no more than one less than the denominator of the fraction. For example, the fraction $\frac{1}{7}$ has a period of 6 (i.e., 7 – 1) because six digits repeat. Many fractions have a repetend far smaller than their denominator, but the limit of the number of digits that can repeat is always one less than the denominator. Often the repetend is much smaller. The repetend for $\frac{1}{9}$, for example, is only one digit

($.\overline{1}$) while it could have been as many as eight, and the repetend for $\frac{1}{11}$ is only two digits ($.\overline{09}$) while it could have been as many as ten. I suppose this is not a major discovery, but I can think of times when it could be useful. Would you like to know the decimal expansion of $\frac{1}{19}$?"

"That's just one of the 999 decimal expansions that you have figured out, isn't it?"

"Yes. You'll be glad to know that the repetend is exactly one less than 19; exactly 18 digits repeat. The fraction $\frac{1}{19} = .\overline{052631578947368421}$."

"And you say you can prove that this is always true for the reciprocal of any whole number? That the period of the repetend is never more than one less than the number?"

"That's right."

"Number theory really is exciting, isn't it?" Bolyai continued. "Let me ask you, though, how long did it take you to find the decimal expansions of the reciprocals of all those whole numbers?"

"Well, I've been working on it for a couple of years, but I have done most of it in the last couple of weeks. I already had a proof that the period would never be greater than one less than the denominator, but I still had to figure out all those numbers because, even if logic tells me it is true, it would take only one number that has a longer period to make me look like a fool. I have no intention of missing anything obvious."

"What else have you been working on, Gauss?"

"Well, I've been exploring least common multiples and greatest common factors for pairs of numbers. If you take the numbers 48 and 64, their greatest common factor is 16 and their least common multiple is 192. When I multiply the greatest common factor and the least common multiple together (16 × 192) I get the same number that I get when I multiply the original numbers: 16 × 192 = 48 × 64. I can prove that for any two numbers, the product of the greatest common factor and the least common multiple is always equal to the product of the two numbers. Give me a pair of composite numbers, Bolyai."

"How about 28 and 42?"

"Okay, their greatest common factor is 14 and their least common multiple is 84. The product of 28 × 42 is 1,176. The product of 14 × 84 is 1,176. Right?"

"And you did that all in your head!" said Bolyai.

"You could have done it too. Remember, though, I've been playing around with this a lot lately," said Gauss.

"So in fact this is a good way to be sure that you have the correct greatest common factor and least common multiple of a pair of numbers," said Bolyai. "If the product of the greatest common factor and the least common multiple is not the same as the product of the two numbers, then you have made a mistake."

"That's right: a self-correcting process," said Gauss.

Gauss and Bolyai became close friends, often walking in the forest together or going out for something to eat after a full day of studying. Over time, they talked about many things, and eventually they decided to formalize their friendship. They went out to a *Kneipe* one evening, interlocked elbows as was the custom when friends made such a commitment, drank a toast to their friendship, and pledged to be friends for life. That was when they agreed to call one another by first name, Carl and Wolfgang, and to use the familiar *Du* form of address for *you* instead of the formal *Sie*. This was a major step in their friendship. Wolfgang was the only friend to whom Carl wrote using the familiar form of *you* in his correspondence.

"Carl, your friendship has been a wonderful part of my studies at Göttingen," said Wolfgang.

"Yes, Wolfgang," agreed Carl, "my studies, too, have been more satisfying because I can share my thoughts with you. I have had friends before, but I couldn't communicate with them on the same level that we communicate."

"Do you remember that day a month or so ago when we went walking in the woods together, each of us working on a difficult problem that we were determined to solve?" asked Wolfgang.

"Yes," said Carl. "I needed to think, but I didn't need to be alone. In a way, we were alone together. Other people would have broken the silence because they would have found it uncomfortable, but we both felt thoroughly comfortable in our silence."

"I made some excellent progress on solving my problem that day," said Wolfgang. "Did you?"

"Oh, yes! I was working on the fundamental theorem of algebra, going back through the steps in Euler's proof. I have to say, I found some gaping holes in it," Carl said.

"Gaping holes in Euler's proof?" gasped Wolfgang.

"Yes, I know it sounds like sacrilege, but I'm sure I'm right," said Carl. "I know Euler was an impressive mathematician, but in this case he made some assumptions that are not valid. I'll show you what I mean sometime."

In 1797, Gauss and Bolyai together walked 60 miles from Göttingen to Braunschweig to visit Gauss' mother. Bolyai was a long way from his own home in Transylvania, and he was naturally curious about his remarkable friend's home and family. It was a long walk, providing lots of time for talk as well as silent contemplation, which both young men enjoyed.

As they walked, Gauss' 17-gon came up. "Wolfgang, I wanted to tell you that if you want to hold onto the paper on which I figured out the 17-gon, you can keep it. I don't need it anymore. Take it back with you to Transylvania when you finish your degree as a remembrance of me."

"I would love to do that, Carl. You are sure you won't need it?"

"No. I have no further use for it."

"You know," Wolfgang continued, "I also still have that drawing you made of Kästner one day in class. Remember how you had him standing at the blackboard with a subtraction problem complete with an arithmetic mistake?" asked Wolfgang.

"I'd forgotten about that!" said Carl.

"I was amused," said Wolfgang, "and I think I'll keep them both together—my keepsakes of our university days together."

When they reached Braunschweig the next day, Gauss' mother greeted her son and his friend with open arms.

"My dearest Carl!"

"*Mutter*! How wonderful to see you!"

"Ach, Carl. You look great. Is this your friend *Herr* Bolyai?"

"Yes, this is Wolfgang. I am glad you are able to meet him at last."

"Yes, *Frau* [Mrs.] Gauss, I am delighted to meet the mother of my esteemed friend."

"Well, you must be tired and thirsty. You should be hungry as well. You've come a long way. Here. Sit down, take off your shoes, put up your feet, and let me bring you something to drink."

"You make me feel so welcome, *Frau* Gauss. You don't know how nice it is to be in a home with a mother, when my own mother is so far away."

"My son's friends are like my own son. I am so glad to have you here."

And so the three of them sat down to a supper of bread and cheese and sausage, and talked about all that had been going on in Braunschweig in Carl's absence as well as what the university students had been doing in Göttingen. Later, when Carl went upstairs to prepare a bed for his friend, Frau Gauss asked Bolyai the question she had been so eager to ask him.

"*Herr* Bolyai, what do you think will ever become of my son Carl? Will he ever do anything important?"

"But *Frau* Gauss, don't you know that he is already Europe's finest mathematician?"

Tears welled up in her eyes when she heard that. She had always known that Carl was brilliant and hard-working, but she was just an ignorant woman. What did she know? But if this nobleman from Hungary who was also studying mathematics with her son described him that way, it must be true. This was a beautiful day indeed!

Later that evening, Carl had a question for his mother: "*Mutter*, what is the exact date of my birthday? I know I was born sometime after Easter in 1777, but I'd like to know the date."

"Well, Carl, let me think. It was a Wednesday, the week before the Ascension, the day that Christ is said to have gone up to heaven."

"Okay, let's see. The Ascension is the 40th day after Easter. It's usually a Thursday—the Thursday of the sixth week after Easter, isn't it?"

"That sounds right."

"So that means I need to know the date of Easter in 1777 if I want to find the exact date of my birth. You know, the way Easter is calculated is very strange, *Mutter*: the first Sunday after the first full moon after the spring equinox. It is a combination of the phases of the moon and the solar calendar. Only the Christian church knows for sure when it's coming. As a mathematician, I would like to know, too."

"Well, Carl, perhaps you can figure out a way to do it. I'm sure I don't know."

As they walked back to Göttingen the next day, Bolyai asked Gauss about his father. "Carl, your father is still alive, isn't he? Do you have any contact with him?"

"Very little," Carl replied. "My father really doesn't understand what I am up to. To him, what I do doesn't qualify as work. Manual labor is the only kind of work he has ever known, and he believes I should feel guilty about not having calluses on my hands or an aching back at the end of the day. He thinks that I need to grow up and get a job."

As they walked, Gauss contemplated how to figure the date of Easter. "Wolfgang, I think the solution has to deal with remainders—what is left after you divide by a number. I think I will have to divide the year by seven since there are seven days in a week, and then find the remainder after that division. I'll have to use 30 also, since on average there are 30 days in a month, and four since there are about four weeks in a month. Hmm. I guess I need to find the dates of Easter in the last few years in order to construct this properly. Yes, I think I should be able to figure this out."

"Wolfgang, I've got it!" Carl greeted his friend a week later.

"You've got what?" asked Wolfgang.

"I've got my birth date. I figured it out. I was born on April 30, 1777," said Carl.

"Did you find an old calendar in the library?" asked Wolfgang.

Number Theory

"Yes, I needed to use several old calendars," said Carl, "but now I have a formula that will allow me to find the date of Easter for any year between 1582 and 2099. It all has to do with those remainders after division that I was talking about. That's how I dealt with the phases of the moon and the sun together."

"Are you willing to share your method with me?" asked Wolfgang.

"Sure, it's not too bad. I've got this chart, and you have to look at it to see how my method works."

Years	M	N
1582–1699	22	2
1700–1799	23	3
1800–1899	23	4
1900–2099	24	5

"Okay," Wolfgang began, "since we are talking about the year 1777, $M = 23$ and $N = 3$. I don't feel any closer to the solution than I was before."

"That's because I haven't given you the formulas yet!" said Carl. "This will take a minute. We are looking for five numbers, and those numbers are the remainders after we have divided. Watch this:

"The year for which I want to find Easter is 1777. First I divide the year 1777 by 4 and it gives me a quotient of 444 with a remainder of 1. That means the remainder $a = 1$. Remember, it really doesn't matter how many times four goes into 1777—all we are interested in is the remainder.

"Next I divide the year 1777 by 7 and that gives me a quotient of 253 with a remainder of 6, so the remainder $b = 6$. Once again, we don't care how many times 7 goes into it—we are interested only in the remainder. Are you with me?"

"I've got it. What's next?" Wolfgang asked.

"Now I divide 1777 by 19 and that gives me a quotient of 93 with a remainder of 10, so $c = 10$," explained Carl.

"Let me be sure I'm following you. What we know now is that $a = 1$, $b = 6$, and $c = 10$. You said there are five numbers," said Wolfgang.

"Gauss continued: "Next we need to find d with this formula using our value for c and M from the table: $d = \frac{19c+M}{30}$. That is going to be $\frac{19 \times 10 + 23}{30}$ and that gives me a quotient of 7 with a remainder of 3, so $d = 3$."

"Okay."

"The last formula," Gauss continued, "is the most complicated. Here is the formula for e: $\frac{2a+4b+6d+N}{7}$ and that gives me $\frac{2+4\times6+6\times3+3}{7} = \frac{47}{7}$ which gives me a quotient of 6 with a remainder of 5, so $e = 5$.

"From here on, it's easy. We'll get the date of Easter by using one of these two formulas: $22 + d + e$ is the date in March or $d + e - 9$ is the date in April. We use whichever formula gives a possible date—like it can't be March 53 since there are only 31 days in March."

"All right," said Wolfgang. "Let me take it from here. $22 + d + e$ gives me $22 + 3 + 5 =$ March 30. That was Easter in 1777. That's not your birthday."

"No. Now we need to go forward 40 days to get to Ascension Day. No, that's not right. In fact, we have to use Easter as day one and count 39 days from there because the calendar is constructed on the assumption that there is no number zero. There is one day left in March after Easter and we have 30 days in April, so Ascension Day must be eight days later on May 8, and that is a Thursday. My mother said I was born on the Wednesday of the week before that, so my birthday should be April 30."

"So once I know M and N for the year I'm looking at, I have to use your formulas—what were those formulas again?" asked Bolyai.

"Here they are:

Year divided by 4	Remainder = a
Year divided by 7	Remainder = b
Year divided by 7	Remainder = c
$\dfrac{19c + M}{30}$	Remainder = d
$\dfrac{2a + 4b + 6d + N}{7}$	Remainder = e
$22 + d + e =$ $d + e - 9 =$	date in March date in April

"All these calculations are focusing on the remainder after division, aren't they?" asked Bolyai.

"Yes, Wolfgang. I have discovered that there are many times when the remainder is the only significant feature of division," Carl explained.

Within a few years, Gauss established this as a new part of number theory, which he called modular arithmetic. He is recognized as the originator of this branch of number theory.

Part III

Gifted Astronomer, Father of a Young Family
1798–1814

Carl Friedrich Gauss, PhD

In 1798 Gauss left the university of Göttingen and returned to Braunschweig. He was now 21 years old, a sturdy young man, a little below average in height (about 5'2"), with blond hair and piercing deep blue eyes. He had not finished his mathematical studies, but he was ready now to pursue them on his own.

When he first returned to Braunschweig, his mother assumed he would move in with them, but Gauss decided against that.

"Carl, we can make room for you here. How will you support yourself?"

"*Mutter*, I'll get along. I need peace and solitude to do my work. You do understand that my studies are not done, don't you? I have finished with Göttingen—I have already benefited from all that Göttingen has to offer. Now I need to work. I'll find a way to get by on my own. Don't worry. I've found a room that's not too expensive, and I have a lead on a tutoring job that would pay pretty well. No, *Mutter*, I cannot do my work at home."

A few weeks later while standing in line at the butcher's shop, some of the women were talking.

"I see Carl Gauss is back in town. He must have finished up at the university. I wonder what he'll do now."

"I heard he's not going to do anything. I heard he's not even looking for work."

"Oh, that couldn't be true! A healthy, strong young man like Carl? No. You must have gotten that wrong."

"Well, I heard that he has no intention of looking for work. I hear he visits the stationer's shop every few days and buys paper, and then he just goes back to his room and writes. I talked to *Frau* Schröder, who rents him his room on Wendenstrasse. She says you wouldn't believe the number of candles he uses! She says that all over his room there are papers and notebooks. She has looked at those papers, and they are covered with lots of strange symbols, and the words are not proper German. She's never seen the likes of it. I hear he's planning to go to Helmstedt for a while too. If he's going to do nothing, I don't see why Braunschweig isn't a good enough place to do it!"

"I have heard that he speaks English and French just as well as you and I speak German."

"But wait! Listen to this: I have heard that he does all his writing in Latin, of all things! When did anyone speak Latin in the last 200 years?"

"What's the point of a man learning to speak English or French or Latin when German is good enough for anyone here in Braunschweig? Can you imagine what his poor mother is going through?"

"Oh, I don't think she's suffering. She thinks her son is a genius, and she is as proud as she can be. She says he's the greatest mathematician in Europe. She says that is what his Hungarian nobleman-friend told her."

"And what will she do when he dies young, the way all geniuses do?"

"She probably hasn't thought that one through."

"And does playing with numbers and scribbling nonsense in Latin put bread on the table? I think not! The Duke has supported him all this time, and I suppose that may have given her the idea that her son is special, but really! Anyone living in modern Braunschweig must work. We don't have room for slackers. I wonder what he's living on?"

"Credit, I hear. Not a very pleasant existence. I'm sure he could get a job somewhere. I wonder why he doesn't try. I guess he just doesn't have any ambition. It's a pity."

The Duke also heard that Carl was back in Braunschweig, and he sent a message to Carl that he should come to the palace for a talk.

"Well, *Herr* Gauss, have you finished your studies at the university?"

"Yes, Your Highness. I think I got all I can from Göttingen and its library, so I have come home to Braunschweig to write. I have some big ideas that I have been working on day and night. I have just about finished two manuscripts. One of them outlines my new proof of the fundamental theorem of algebra. The other explains some of the new things I have come up with in number theory and proves the fundamental theorem of arithmetic. I've gone to Helmstedt and talked my ideas over with Professor Pfaff, and he encouraged me to take a room with him and continue my work there for a while. I think I may do that. I think he could help me clarify my thoughts and streamline my reasoning. Professor Pfaff is a fine mathematician. Your Highness, I have so many things that I want to get down on paper!"

"What are you living on, *Herr* Gauss?"

"Well, that is a bit difficult. I thought I might be able to get a position tutoring that Russian nobleman's daughter here, but a Frenchman beat me to that. Oh, well. I'm sure I can figure out some way to support myself."

"Why don't I help for a bit longer?"

"I hate to impose on you anymore, sir."

"It sounds as if you are doing important things. How would you feel about submitting a thesis for a PhD at Helmstedt under Pfaff?"

"Well, I could use my proof of the fundamental theorem of algebra for that, Your Highness."

"Excellent. I'll tell him of our plan. I'd like you to get a doctorate. That would legitimize your status as a mathematician."

Now that the Duke was once more helping him financially, Gauss no longer had to worry about how he would pay for his paper and ink or where he would find the money for food or rent. He realized that he was a very lucky young man. He didn't have to spend time developing engines of war for his king the way that Archimedes did. He didn't have to tutor German princesses the way that Euler did. Only three or four people in all of Europe were able to practice the art of mathematics as a fulltime career.

The Duke had put him in a position to do mathematics with rare intensity. The Duke knew that he could trust Gauss' genius to lead him toward significant work. Gauss, a true scholar, was peculiarly suited for this role. Never doubting his own ability, he attacked each new challenge vigorously and kept working on it until he was satisfied with the results.

"If others would think about mathematical truths as deeply and as continuously as I have, they would make my discoveries," he commented when he was older. In general, those who do great mathematics spend vast and deep and constant energy and attention on it.

One day the next week, he happened to encounter his father in town. His father had heard the gossip, and he was glad to have the chance to set Carl straight: "Well, Carl, I hear the Duke is giving you more money. What do you need it for? Our family has never taken charity. I think you should know the satisfaction of doing a hard day's work and getting paid for it."

"Father, you don't understand," Carl answered. "I work very hard, but I work with my brain rather than my hands. You will see. I expect to do some important work."

"You call that work?" his father asked. "I don't call that work. Grow up, boy! I'm tired of making excuses for you."

"I'm sorry, Father. Mathematics is the foundation of science, and my challenge is to explore it in important ways. I don't expect you to understand that, but you should know that I have not asked the Duke for his support. He has offered it to me because he believes in my work. He believes that it is important," Carl explained.

"That's nonsense! Goodbye, Carl." Carl sadly watched his father stride away. He and his father never spoke to each other again.

A few months later, he wrote to his friend Bolyai, who was still in Göttingen:

> I finally have some real news for you, Wolfgang. I have submitted a thesis proving the fundamental theorem of algebra to Professor Pfaff at the university at Helmstedt and the committee has accepted it! Not only that, the Duke will pay for publishing it.
>
> Over the past few months I have spent some time in Helmstedt working with Professor Pfaff. He really is a fine mathematician as well as a kind and interesting man. In some ways he seems almost childlike in his innocence and openness, but then he brings me back to reality with his penetrating insights into mathematics. You remember how we struggled with Kästner's incompetence in Göttingen? I told you then that I had heard that Pfaff was a fine scholar, and it is true. He seems to respect my work, and with his influence they have awarded me a doctor's degree without requiring me to defend the thesis. Apparently that doesn't usually happen. They seem to believe that my thesis stands on its own—that my arguments are valid. What a relief!

Years ago when Martin Bartels and I were working our way through algebra, I wondered how we could be sure the fundamental theorem of algebra is true. Martin was quite short with me at the time, observing that I could not reinvent all of mathematics on my own. However, I believed then as now that some parts of mathematics need reinventing. In my thesis, the first thing I do is to explain where Euler went wrong in his proof of the theorem, and then I go on to d'Alembert's and Lagrange's proofs, demonstrating their mistakes as well. I may be only 22 years old, but I have spent the past ten years exploring this theorem in detail. I was a little worried at first that Pfaff might think I was too bold, attacking the major mathematicians of the last 100 years, but I have convinced him. He agrees that I am right, and so you may now call me Dr. Gauss!

Next I need to finish up my manuscript on number theory—the most beautiful part of mathematics. You may have heard me call number theory the Queen of Mathematics—I still believe it! This manuscript includes many of my discoveries of the last four years. With a doctorate, I believe my new work will be taken seriously. After that, I have so many more ideas that I need to develop—the difficulty is finding the time to pursue them. I understand there are people who are bored. I can't imagine that!

I hope things are going well with your work.
Your friend,
Carl

Eighteen months later when Gauss completed *Discourses on Number Theory*, in which he proved the fundamental theorem of arithmetic, the Duke paid for its publication. Gauss dedicated the work to him: "I dedicate this work to you, Karl Wilhelm Ferdinand, Duke of Braunschweig. Your understanding and support over the last 13 years has made it possible. You believed in me and you enabled me to devote these years to the fruitful study of mathematics. The life of a scholar is the life that I have chosen, and you have graciously removed all the obstacles that would normally have made it impossible. I thank you heartily."

Although it may have appeared to others that Gauss had an easy life in Braunschweig and Helmstedt, his schedule was demanding. He was always finding more ideas that needed to be developed, and there was never enough time for all of them. In these few years, he discovered on his own virtually all the mathematics that all the mathematicians in Europe would produce in the entire century that followed, although he published little of it. Expressing the major concepts he wanted to explain was a grueling task, always demanding more and more of his genius.

When Bolyai was reaching the end of his studies in Göttingen, he and Gauss agreed to meet once more in Clausthal, a village in the Hartz Mountains between Göttingen and Braunschweig. They met at an inn in the village, and as they sat in the pleasant common room sipping beer together in front of the warm, tiled stove, they had much to talk about.

"Carl, you have accomplished many things in the time I have known you," said Bolyai.

"You, too, have accomplished a great deal, Wolfgang," said Carl.

"But Carl, you have completed your doctorate and your book on number theory," said Bolyai.

"I have been very fortunate," said Gauss. "When I was a child, I dreamed of many things. I wanted very much to meet the Duke, and I was fortunate to do that when I was 11 years old. I wanted to study mathematics seriously, and I am doing that thanks to the Duke. My third wish was to be rich, but I don't seem to have gotten too far in that."

"Do you go to bed hungry?" asked Bolyai.

"No, I have enough to eat," Carl admitted. "The Duke has supported me for many years. I have noticed that having enough money makes everything else easier. When I was growing up, my family never had quite enough, and it was always a struggle. That probably is part of the reason that my father was always so disagreeable. I would like to have enough so that I am not constantly worried about basic necessities. You are right that being rich is not important, but having enough allows us to reach for higher things."

"That is true. Let's hope that both of us always live in relative comfort. Let's also hope that both of us thrive as scholars. Carl, I would like to propose a ceremonial smoking of pipes," said Bolyai.

"Why don't we exchange pipes? You could take mine and I could take yours," suggested Gauss.

"I like that idea. If I smoked your pipe at sundown each evening, could you do the same with mine?" asked Bolyai.

"Yes," agreed Gauss. "Then even though we would be far apart, we would have a common experience each afternoon. Our friendship does not need to end just because we live in different places."

Although they never saw each other again after that, they corresponded faithfully for 50 years.

The Planetoid Ceres

In 1801, Gauss read about a tantalizing discovery by an Italian astronomer named Giuseppi Piazzi. Piazzi, having discovered a small planet that he called Ceres, made careful notes of his 24 nights of sightings. Unfortunately, they allowed a glimpse of only 9° (out of a total of 360°) of its orbit before it disappeared from his sight. His findings were published in the *Monthly Correspondence of Mapping the Earth and the Sky* (a journal edited at Gotha by the astronomer Franz Xaver von Zach) for amateur and professional astronomers.

Gauss found the sightings an irresistible challenge. This was a mathematical problem that was easily within his amazing calculating abilities, and, using his method of least squares, he soon had a formula for the orbit and was able to predict when and where in the night sky it would be visible in Germany and elsewhere in Europe over the next several years. Although several other astronomers had been working on it too, they limited themselves by starting with an assumption about what the orbit was and then trying to make the data fit the assumption. No matter what they assumed, the orbit didn't fit. Because Gauss' only assumption was that it had to be an ellipse of some sort, his only challenge was to find the ellipse that matched the 24 sightings. Once he discovered the formula, he sent it to several astronomers and his prediction was announced to the astronomical community in the *Monthly Correspondence*. Although the weather made it impossible to confirm his figures immediately, when the skies finally cleared, Ceres was rediscovered in December 1801, exactly where Gauss' formula had predicted! Gauss was an instant celebrity. He was immediately elected a corresponding member of the St. Petersburg Academy of Sciences and invited to come to St. Petersburg, the capital of Russia.

When Gauss read in the newspaper about Wilhelm Olbers' rediscovery of Ceres in January 1802, he wrote directly to Olbers to get his new observations. A friendship soon developed between Gauss and Olbers, a physician who lived in Bremen, practicing medicine on the lower floor of his house during the day, working seriously at astronomy in his attic observatory at night, and never sleeping in the middle level for more than four out of every 24 hours. Olbers, who had studied both astronomy and medicine at Göttingen 20 years before Gauss, had published a method for calculating the orbit of a comet in 1779. As a subscriber to the *Monthly Correspondence*, Gauss had read about Olbers long before he met him. Gauss traveled the 200 miles north to visit Olbers in Bremen several times during his life.

On March 28, 1802, Olbers discovered a new object in the heavens. Since it did not appear in any of his star charts, he noted its precise location. The next night he looked again, and, sure enough, it was not in exactly the same spot. That encouraged his suspicion that it was a planet, a word that comes from the

Greek word for wanderer, since planets appear to wander in the night sky. The following night Olbers observed again, and once more it appeared to have moved even farther. The next day he consulted with an astronomer friend in Bremen. He and the friend both observed that night, and once again they found that it had moved.

A week later it was time to consult with young Gauss. Olbers sent Gauss his observations and, using his newly developed techniques, Gauss was able to figure the orbit of the new planet, Pallas, easily. He had the new planet's orbit calculated by April 18, only 20 days after Olbers' initial sighting.

His friendship with Olbers helped Gauss both professionally and personally. At the professional level, the older man introduced Carl to several other astronomers in Bremen.

"*Herr* Doctor Gauss," observed one of them, "all of us have been working for years on calculating orbits of planets, but you do it so quickly. How do you do it?"

"Yes," observed another, "it took Euler three days to calculate the orbit of a planet, yet you do it in three hours."

Gauss couldn't resist this rejoinder: "If I had spent three days doing such calculating, I too would have gone blind," a cruel allusion to Euler's many years of blindness. "My calculations are really nothing more than a confirmation of Newton's law of gravity," he added.

Olbers decided that it was important to help Gauss get a good position in astronomy in Berlin, and he worked behind the scenes to help. He saw his work as more urgent when he learned that St. Petersburg in Russia had offered Gauss the directorship of its observatory. Although Gauss was happy to let Olbers work on a position in Berlin for him, in fact he never accepted it. Nor did he accept the position at St. Petersburg.

The Duke heard of Gauss' accomplishment, and he sent a message summoning Gauss for a meeting at the palace the next afternoon. "Now, *Herr* Doctor," the Duke began after first congratulating him on his accomplishment, "I hear you have been offered a position in St. Petersburg."

"Yes, Your Highness, they have offered it, but I don't particularly want to go. My roots are here in Braunschweig, although I must admit the salary is tempting. You have supported me generously for many years."

"It has been a great honor, *Herr* Doctor. You have done well. How would it be if I increased your stipend from 158 Thalers per year to 400 Thalers? Would that make it easier for you to turn down the offer from St. Petersburg?"

"Yes, sir! I had no idea that such an increase was possible. Thank you very much!"

"I knew from the first time Professor Zimmermann introduced you to me that you were destined to become a great scholar. The astronomers of Europe believe that you have joined their ranks as one of their finest. It is a pleasure to support you in this. Never let it be said that Ferdinand, Duke of Braunschweig, didn't do his part to underwrite the advancement of science! My plan is to erect a first-class observatory here in Braunschweig for your further research."

"I will never forget your kind and generous support, Your Highness. I will make sure that your name is always remembered as a friend to astronomy and mathematics."

Some of the residents of Braunschweig were less enthusiastic. "It's obvious that no corn grown on the planet Ceres will ever find its way into the Braunschweig market on Saturday afternoon!"

"That's right! Why does the Duke keep giving more money to young Gauss when he refuses to grow up and get a job? I just don't understand it."

"Neither do I, but I hear his mother is delighted. Her son, the genius, is thriving. My sons would enjoy a life of leisure at the Duke's expense too! It's outrageous!"

The residents of Braunschweig were not the only ones critical of a pure mathematician. A crank wrote to the editor of the *Monthly Correspondence*, complaining that it certainly would be nicer if our scientists did something useful instead of all this theoretical nonsense. Olbers responded in the journal with a poem in French, which Gauss copied out into his own notebook:

> Millions of people on this earth
> Think they know a scientist's worth.
> They cannot grasp it, so they are quick to say
> That no one will use these ideas anyway.
> But we must be charitable, honest and polite,
> For in the end they will realize that science is right.

People who thought they were clever but had no understanding of science and the working of scientists amused Gauss. Abstract science and mathematical research usually work out in the end to have practical applications. The scientists themselves may not dream of these applications, but the theoretical foundations must come first.

Gauss realized that his work on astronomy was limited by his total ignorance of practical observation. He was fortunate that other astronomers were providing him with their observations, but he really should know how to get the data himself. To remedy his limitation, he borrowed a sextant, a clock, and a telescope from Zach, an astronomer whom Olbers had introduced him to.

"*Herr* Doctor Gauss," began Zach, "I am pleased to loan you these tools of the astronomer's trade, but I have one concern: your eyes. You are near-sighted, and I fear that will limit you in using a sextant."

"Thank you so much for the loan of your equipment," responded Gauss. "You may be right that my eyes will restrict my use of the sextant, but I would like to try anyway."

"I hope you prove me wrong," said Zach.

In fact, Gauss' near-sightedness was not a problem. Zach might have had trouble using the sextant with Gauss' eyes, but Gauss knew his own eyes well, and he had been accommodating his nearsightedness for many years. In fact, since he was dealing with great distances, the limitations of his nearsightedness were minimal. We all have limitations on how far we can see, and Gauss' limitations were only slightly greater than Zach's. Gauss' genius more than made up for his myopia.

Gauss' telescope

Gauss began to dream of working out the triangulation of Braunschweig, the necessary first step for an accurate map. Triangulation is based on the fact that a triangle with sides of given lengths is a rigid structure. Any other polygon can be wiggled into a variety of shapes. The triangulation necessary for a map involves measuring the angles formed between several pairs of points that always yield the same location. By 1803 Gauss was working on the measurements necessary to get the exact longitude measures of Braunschweig, Helmstedt, and Wolfenbüttel. Exact measurements of latitude had long been possible. Longitude was another matter, because it had to be measured in relation to the precise time of day or night. The clock that Gauss had borrowed from Zach was not accurate enough for a precise longitude measure, but with it he was able to make a start on the process.

In 1804 Gauss met his friend Olbers at Spa Rehburg, not far from Göttingen, where Olbers was taking a short vacation.

Together they measured the latitude and longitude of a small hill just outside of town. The camaraderie as they worked and talked together was a delight to both of them.

A Wife and a Child

In 1803 Gauss wrote a letter to Bolyai, describing his life in Braunschweig and the people he encountered. Gauss' godfather, Georg Karl Ritter, had taken an interest in his godson over the years. When Gauss returned to Braunschweig, he occasionally took part in social evenings at his godfather's home. He enjoyed these convivial evenings of fellowship, and one of the people he met there was a delightful girl named Johanna Osthoff.

Gauss wrote to Bolyai, "I can hardly believe my good fortune in meeting this remarkable young woman. She is beautiful, wholesome, and bright. She doesn't put on airs; she is just a natural, happy, well-educated young woman. However, I will not tell her of my admiration until I know her better. I certainly would not want to force myself on her. If she can't love me, it would be torture to ask her to share my life. However, I will continue to hope."

For a year, Gauss waited, seeing her when and where he could but making no advances. Finally, when he could wait no longer, he wrote this letter to her: "My very dear friend: I have not found a time to approach you until now, and before I tell you about my desires I want to tell you how much I admire you. You are beautiful and charming. As far as I can tell, you have no flaws. I would like to ask you to be my bride if you could love me too. Since I first met you, my heart has belonged to you. I am not rich, and I don't have very much to offer you, but I believe I have enough for two modest people to live on. If you are willing to marry me, I would be true to you and I would do my very best to support you and love you forever. Could you consider my proposal? Please tell me as soon as you can."

It was three weeks before Johanna replied. While she was flattered by Gauss' proposal, she was humble enough to wonder if she was good enough for him. She was well aware of his reputation as a genius. She and her friend Luise talked it over.

"Luise," Johanna confided to her friend as they sat sewing together one afternoon, "did I tell you that *Herr* Gauss has asked me to marry him. I don't know what to do. What do you think?"

"Well, Johanna, he certainly is impressive. People say he is a genius."

"Do you think I am intelligent enough for him?"

"Yes, of course you are. You may not do the great scientific work he does, but he won't expect that of you. And besides, he can't do mathematics all the time."

"Yes, but someone in the market said that he is engaged to another woman—a woman from a very rich family who has a splendid education. She would certainly be a better wife for *Herr* Gauss than I would. I'm just a common peasant."

"No, you are not, Johanna Osthoff! Do you think I would associate with a common peasant? No. You are the noblest of the noble. *Herr* Gauss couldn't do better than to marry you. Do you think you could love him?"

"Oh, yes! He is so idealistic and so precise in all his actions, and his letter to me is genuinely moving. I just don't know. What if he is already engaged to this other woman? I'd feel like such a fool."

"But Johanna, what do you care about rumors around town? You know that there are people who resent the fact that the Duke has been supporting him all this time. They think *Herr* Gauss should get a job like everyone else. I can imagine some of them manufacturing a rumor to make him look more arrogant. Besides, don't forget that he wrote you that letter."

"Yes, Luise, you are right. I guess some of the townspeople do resent him. I certainly do not. I think the Duke is wise to support him. I guess the Duke wouldn't do it unless he had good reason to think *Herr* Gauss is doing important work. *Herr* Gauss is certainly brilliant, and I think he is going to be a great man. I guess I shouldn't worry about ugly rumors. And, if *Herr* Gauss is engaged to someone else, let him tell me so himself! At least I owe him the courtesy of a reply."

Johanna Osthoff and Carl Friedrich Gauss became engaged on November 22, 1804. The fictional "other woman" never materialized—she was probably the product of idle tongues and jealous thoughts. Carl and Johanna were married almost a year later on October 9, 1805. Carl was 28 years old and Johanna was 25.

Gauss described his elation in a letter to Bolyai: "Three days ago I married the most wonderful woman who ever lived. She is beautiful, and amazingly enough she loves me too. Every time

I look at her I am struck again by her shy and genuine manner, by her innocence and honesty, and by how lucky I am to have married such an angel. I never expected to find such happiness."

They rented a small apartment above a dry goods shop. From the front windows Johanna could look down and see what was happening on the main street in town, and Carl had a table by another window where he could spread out his papers. Johanna happily worked on her sewing projects, preparing for the children who would come before long. They were perfectly happy together.

Johanna was not a great intellect with whom Gauss could discuss his work. She didn't understand his mathematics. She was intelligent and fun, however, and slightly mischievous.

"Carl, where do you suppose my laundry list is?"

"I don't know. Did I have it?" asked Gauss.

"Yes, I think so," said Johanna. "You were working on a drawing when I was making the list, and then I had to run to the butcher's shop. I fear that your ellipses may have some interesting new features—perhaps some arm holes or some delicate lace! Do you mind if I look through these papers?"

"Of course not! Don't worry about the order of the papers. I can easily sort them out again."

"Here it is: between the sketch and the calculations for an orbit. Do you suppose the laundry would get cleaner if we sent it into orbit? Then again, maybe it would take too long—we'd run out of clothes before a year was up."

She provided a delightful balance to his exacting work, understanding his need to concentrate deeply on his work and never expressing jealously over the demanding role his research played

A Wife and a Child

in his life. She was literate and happy, and he considered himself the luckiest man on earth to be married to such a woman.

Johanna and her friend Luise often spent an hour or two together in the afternoon. "Luise, can I pour you another cup of tea?" asked Johanna one afternoon as she and her friend sat down for a visit.

"Yes, that would be lovely, Johanna," said Luise. "This tablecloth is beautiful! Didn't Gauss' uncle, *Herr* Benze, make it?"

"Yes, he makes exquisite damask."

"I wonder how he learned to make it. No one else in Braunschweig is able to make anything like it," said Luise.

"Carl told me his uncle figured it out himself. As a boy, he was apprenticed to a weaver to learn the basic processes, but when he saw some fine damask he decided to try it himself. He devised his own very clever technique, and the results are marvelous. And you know, Luise, he is a wonderful man. He and Carl have always been very close—they spent a lot of time together when Carl was young. When he comes over in the evening, we all have such fun!"

"And you have some lovely examples of his work. What a pleasure it is to have fine things! My mother has ordered some pieces from him for my hope chest."

In 1806, the Gausses' first son, Josef, was born. Josef's name came from Giuseppi (Italian for Joseph) Piazzi, the discoverer of the planetoid Ceres—the planet that first allowed Gauss to show the world his amazing skill as a mathematician and astronomer. Josef's name was the beginning of an astronomical tradition in the Gauss family, as successive children were named after other astronomers. Johanna was a wonderful young mother, and she watched with pleasure as their baby grew. It seemed that their happiness was complete.

"Carl, just look at his darling little hand, perfect in every respect but so tiny! Do you suppose you had tiny hands like that when you were first born?"

"I suppose so. A baby isn't just a miniature adult, though, is he?"

"No, but I can't wait to see how Josef grows up. He is such a fine, strong baby. I wonder what he'll do with his life. Would you like him to grow up to be a mathematician like you?"

"Well, I don't know. We'll have to see what happens."

The Duke and St. Petersburg

The Duke, with some assistance from Zach, worked on plans for the construction of an observatory at Braunschweig with Gauss as its director. When Gauss heard of a fine lens that was available and suitable for a good telescope, he told the Duke about it. The Duke immediately made the money available for Gauss to purchase it. The lens needed some modifications, and it took considerably more money and time than the initial estimate, but once more the Duke provided the money that was needed. Gauss anticipated with pleasure having a real observatory of his own.

However, outside the small-town setting of Braunschweig, Napoleon was aggressively conquering what would become a united Germany 50 years later. Duke Ferdinand of Braunschweig, who was 70 years old at this time, was called up to serve the neighboring king in Prussia. First the king sent him to St. Petersburg to see if he could arrange for the Russians to help in the defense of Prussia, Hannover, and Braunschweig against Napoleon, but the Duke met with nothing but frustration.

Although the Russians were unwilling to commit either men or arms to aid the Prussians, they did have one consuming interest: "Tell us more about this young man Gauss, Your Highness. We have offered him a fine position, but it seems that you have made a counter-offer, which he appears to have accepted. Surely he would be better off here in our institute! Scholars need to work with other scholars in an environment conducive to scholarly research. Your own *Herr* Leibniz made the plans for our institute, and Euler spent many productive years working here. It is the ideal setting for a scholar of Gauss' caliber. Won't you reconsider, Your Highness? It would be a tremendous contribution to the intellectual world."

When the Duke returned to Braunschweig, he increased Gauss' salary by another 200 *Thalers* to the impressive sum of 600 *Thalers* per year. Gauss learned of the increase on his 29th birthday, April 30, 1806. Gauss was a Braunschweig man, and he would remain so as long as the Duke of Braunschweig had anything to say about it.

As Napoleon's forces continued their invasion, the aged Duke was called to serve as general in charge of the Prussian forces, and he was mortally wounded in the battle at Auerstedt, near Jena, on October 14, 1806. His quickly assembled and poorly trained army was vastly outnumbered. After the battle, despite entreaties that the Duke be allowed to return to his home to die, Napoleon refused, and Gauss looked out his window early one morning to see a long carriage transporting his beloved Duke to die in free territory northwest of Hamburg. It was a sad day for Braunschweig and for Gauss. The Duke had been Gauss' kind benefactor for 15 years. He was also the sole source of his income, and that privilege died with the Duke. His recent salary increase became meaningless. Suddenly, Gauss needed to find a way to support himself and his family. He needed a job.

As Napoleon's troops occupied Braunschweig, the commander in charge received a curious order from Paris: Make sure that the great mathematician Gauss is not injured. The death of the Duke hurt Gauss both personally and financially, but that was no worry to the French. However, the life of Gauss the scholar was.

Napoleon, as a well-educated man, knew the story of Archimedes, who was the victim of a senseless murder when the Roman general Marcellus was conquering Syracuse. Although the order to spare Archimedes had been clear, a common soldier, who found him at his drawings, ordered Archimedes to come with him to the general immediately. Because Archimedes wanted to finish his drawing first, the soldier killed him on the spot. Napoleon, the modern and enlightened ruler, would not have such a blot on his record! No! Gauss was to be spared for the future of mathematics.

For the past few years, Gauss had been corresponding with a mathematician in France who identified himself as *Monsieur* LeBlanc. M. LeBlanc had studied Gauss' work *Discourses on Number Theory*, and had asked some penetrating questions about certain sections. Clearly he had read the work with greater understanding than most. It turned out that M. LeBlanc was in fact a woman named Sophie Germain, one of the finest mathematicians and scientists in France. She had feared that Gauss would not take her correspondence seriously if he knew that she was a woman. However, when she realized that his life might be in danger with Napoleon's conquering of Braunschweig, she prevailed upon Napoleon and his General Pernety, using her own name, to protect Gauss. When General Pernety told Gauss that he had been sent by *Mlle.* Sophie Germain of Paris, Gauss was stunned. He didn't know a woman in Paris named Sophie Germain!

Three months later, when she wrote to Gauss and told him the whole story, he replied with enthusiasm: "Madam: I can't tell you how delighted I was to get your letter. I appreciate your concern for my safety, although I have never been in real danger. But when I learned that the illustrious M. LeBlanc is in fact a woman, I was overjoyed. We make it so difficult for a woman to excel in mathematics, but despite these difficulties you have outdistanced many male mathematicians in your accomplishments. You have a rare talent. I already respected you as a mathematician, but now I respect you doubly for having been able to accomplish so much, given the discrimination a woman faces in the fields of science."

However, protecting Gauss' life did not stop Napoleon's troops later from demanding a fine of 2,000 *francs* from Gauss. Napoleon needed to pay for this very expensive war, and those whom he had defeated were the logical source of funds. It was impossible for Gauss to find so much money. Although his friend Olbers (who was a very wealthy man) in Bremen, mathematicians Sophie Germain and Laplace in France, and other friends tried to help him, he was too proud to accept their help. In the end, when an anonymous donor paid the fee, Gauss could not refuse it since there was no return address.

Shortly after the Duke's death, two things happened concerning Gauss' universities. The first was that Napoleon's forces determined that the university at Helmstedt was redundant and closed it. This meant that the Duke of Braunschweig's university that had granted Gauss his doctorate was no longer extant. His advisor, Pfaff, moved to the university at Halle in eastern Germany and soon became the director of its observatory. The second event was that the University of Göttingen, where Gauss had studied for three years, offered him a professorship as director of its observatory. Gauss was visiting Olbers in Bremen when the offer from Göttingen came.

"*Herr* Olbers, I have been called to serve as director of the observatory at Göttingen. What do you think? Should I accept the offer?"

"Without a doubt, *Herr* Doctor," said Olbers. "This is a magnificent opportunity for you." In fact, Olbers was responsible in great part for the offer from Göttingen although Gauss didn't know it at the time.

"Yes, I think it is, but teaching is not particularly appealing to me. Yes, yes, I know I must be realistic. I was fortunate that the Duke supported my research all these years, and I cannot expect to find another generous benefactor. I know that. I also know that the observatory position would involve far less teaching than a professorship in pure mathematics. That would require a tremendous amount of lecturing. You are right. The observatory looks good, and I do need a way to support my family. As a bachelor, I could limp along with few belongings and no definite prospects for the future. When I married Johanna, I left that life behind, and I am glad of it. Without her, my life wouldn't be worth living. Yes, *Herr* Olbers, I think I must accept the position."

The family moved to Göttingen, where they rented a large apartment on the second floor of a building close to the university, only a few meters from the old observatory. It was comfortable, and the Gausses lived there for the next eight years until the new observatory was completed. Gauss lived in Göttingen for the rest of his life.

"You know, Johanna," Gauss commented to his wife after they had been living in Göttingen for more than a year, "I am beginning to think that the best way to learn mathematics is through private study. That is the way I learned it, and the result seems more than adequate. I think I would prefer simply to give my students a text and let them come to me when they encounter problems. That is the way I worked with Zimmermann at the college."

"But Carl," Johanna objected, "don't you think it's true that other people need more structure to their learning than you do?"

"I suppose that's possible. Of the three students I am teaching this winter, one is modestly prepared, one is less than modestly prepared, and the third has neither ability nor background. That seems to be what I can expect in teaching. You are right. Few of my students would be able to make any progress in mathematics working on their own."

Over the years Gauss taught a limited number of courses, lecturing on the same topics over and over again. However, if he was going to teach, and he had no choice about that, the teaching had to be brilliant. He was a perfectionist in all things. Thus, he made the burden on himself greater at the same time that he reached only a small group of students.

At that time, Göttingen was considered one of the finest German universities. The great mathematician Möbius (the topologist who conceived of the Möbius strip with its single surface) was one of Gauss' most talented students. Despite his complaints, Gauss did have other worthy students over time, and over the years some of them became his closest friends and colleagues.

Richard Dedekind, another great mathematician who studied under Gauss, wrote this description of a series of lectures he felt privileged to have attended: "The lecture hall was small, with a table surrounded by chairs. There were nine of us in the class, and in general every one of us was there every time. It was always a little cramped, and the last people to arrive would have to sit very close to Gauss and take notes in their laps. Gauss spoke clearly and simply, occasionally writing an important word on the blackboard in his beautiful handwriting. If he needed to provide us with a specific number, he always had that number on a slip of paper, which he withdrew from his pocket when he needed it. At the end of the course, he complimented us on our regular attendance and apologized for the fact that it probably had been rather dry."

"Carl, you really don't enjoy the teaching, do you?" asked Johanna one evening after she put Josef to bed.

"No, I don't. However, I don't see any better prospects elsewhere. This is a fine university, I will have a wonderful observa-

Sample of Gauss' handwriting: "Observatory in Göttingen"

tory here, and I don't have to teach all day every day. I do have lots of time to devote to my own work. It's not so bad, Johanna."

"Sometimes I think that you accepted this position just because of Josef and me. I know you feel responsible, but I almost wish you had waited a bit longer to see if you could find a pure research position somewhere. I think people really do consider you the finest mathematician and astronomer in all of Europe, Carl."

"My research has gone well. I have made some great discoveries, although none of them more impressive than my first discovery of the 17-gon. Now that was a brilliant stroke, if I do say so myself!"

"But you need to keep at it as much as you can. Who knows how many other brilliant ideas will hit you if you give them more time!"

"Actually, at this point in my career, having a regular schedule may be a good thing. Many people suggest that I should devote myself to pure mathematics. That certainly is what Pfaff, my doctor-father, thinks! But in fact I get real pleasure out of my astronomy, and it forces me to think mathematically in new ways.

Newton wouldn't have been half the mathematician that he was if he hadn't been an astronomer as well. No, my little *Hännchen*, I think this position at the university and its observatory are perfect for my professional life at the same time that they allow me to support our growing family."

"You care too much for us and too little for yourself, Carl, but I guess your mind is made up. I have to admit I do appreciate the steady income."

In December 1807, Johanna's mother sent a package with some sausage and Christmas presents for little Josef, who was sick. "Carl, look at what my mother has sent to Josef! A table and chair just his size. He will certainly love them when he is feeling better."

"Yes, I hope that happens soon—I mean that he starts feeling better. I hate to see a sick baby. He has been so quiet these last few days. I'll be relieved to hear him let out a healthy cry once more!"

"My mother doesn't know how sick he is, but I do think he is starting to perk up again. I think he is beginning to recover."

"I hope you're right, *Hännchen*. I hope you're right."

Josef did recover, much to everyone's relief, and soon he was playing happily with the little table and chair.

Professor of Astronomy

On February 29, 1808, the Gausses' first daughter, Wilhelmine, was born. Gauss jokingly lamented in a letter that it was a pity she would have only one birthday every four years on the 29th of February! She was named after Gauss' astronomer friend Wilhelm Olbers, and they called her Minna. She was a very big baby. The clothes Johanna had made for Josef to wear in his first few months were already too small for her the day she was born! It wasn't until three weeks after Minna's birth that Johanna was able to get up and care for her family. Childbirth had been hard on her. Today there would have been worry that the mother of such a big baby might have diabetes. However, little Minna was a joy in every way, and the young family was happy.

"Carl, you'll never guess what Josef did today! He picked up one of your books, climbed up into a chair, and sat there as if he were reading it! All he needed was a little velvet cap and the picture would have been complete! I wonder what kinds of things you were doing at his age."

"Well, according to my mother, I was a prodigy from the very beginning, but I suspect she has improved on the stories over time. However, I wouldn't push Josef. Not everyone is possessed by mathematics as I have always been. And it's probably a good thing. I hope Josef will be able to do something using the new technology. There are so many exciting fields just ready to open up, and I'd like to think he has real potential. I hear there is a possibility that some day steam power will provide energy for factories, a far more reliable source of energy than water power, which is so dependent on the season and the weather. I hear that in America Mr. Fulton is propelling boats with steam power. Can you imagine that? Before you know it, they may even find some way to propel carriages on land! Maybe Joseph will be able to get in on that excitement."

"Oh, Carl! Did I mention that Minna got her first tooth? She hadn't even been fussy. This morning I thought I saw something in her mouth and I felt gently with my finger, and sure enough there was a tooth! Amazingly, it is a top tooth. I thought children always got their bottom teeth first. She's only nine months old, so this is pretty young. Josef didn't have a tooth in his head on his first birthday. And Carl, doesn't she seem like an unusually bright baby to you?"

"It's probably a little early to make predictions," said Gauss, "but she certainly seems to be alert. She also seems to be devoted to her big brother."

"Yes, and he is always kind to her. I don't sense any jealousy between them."

"I certainly haven't ever seen any sign of it."

Suddenly the conversation turned more somber. "Johanna, did I tell you that I got word today that my father has died?" Carl said to his wife.

"Oh, no, Carl," Johanna said. "I'm so sorry!"

"Thank you, but it really is not so hard on me. I have never been close to him. He never understood me, and I suppose I never understood him either. He was a good man, but very rough. I think my mother's life may be a little easier now. I never saw any tenderness between them."

The Gausses often entertained scholars from all over Europe in their home. One Professor Bode, a rather pompous man, was regaling the Gauss family with tales of a parrot he had encountered when he was traveling in Massachusetts. The parrot, named Socrates, would answer questions in Greek! Gauss immediately confided that he had a little pet finch that he had also taught to talk, but not in Greek—only in the Braunschweig dialect of German. However, Gauss had to admit that this bird was pretty clever too. Just a few days before, Gauss had asked the bird if he should smoke a cigar or his pipe, and the bird had responded after just a short pause, *"Piep!"*

As Gauss reestablished his contacts with people at the university, there were some surprises. As a student he had attended lectures that were taught by those whom he now called colleagues. He came to realize that Kästner, who had died in 1800, had not been the fool he had thought. In fact, the man had been helpful to him even if he was not a captivating lecturer. Gauss considered that perhaps he too could have fallen into the trap of ponderous lecturing if he faced nothing but class after class of unmotivated students.

As a professor and his wife in Göttingen, the Gausses were occasionally invited out to a social evening of dancing. Johanna enjoyed these activities, but she was not so enthusiastic about afternoon games of cards with the other faculty wives. She told Gauss that she preferred to stay home and play with the children. However, she was not a loner. She had friends with whom she shared pleasant afternoons.

Johanna's friend Minna Waldeck, the daughter of a law professor at the university, often spent the afternoon with Johanna and the children. She was eight years younger than Johanna, but the two shared many interests. While the women sewed and knitted together, the children played happily around them.

"*Fräulein* [Miss] Waldeck," said Josef one afternoon, "would you like to go out riding with me? My hobby horse and I are headed for the country."

"But Josef, I didn't bring a horse with me. How can I go riding with you if I don't have my horse?"

"Oh, don't worry. I have an extra horse. It is in the box in the nursery. Whooooa, boy! I'll be right back. Here is a horse for you. His name is Rossi."

"What is your horse's name?"

"This is Gallup. Gallup and I go out riding every day and we have lots of fun together, don't we, Gallup? Shall we head toward Braunschweig, *Fräulein* Waldeck?"

"Yes, but Josef, I have to be back in time for supper."

"So do I! We'll see you later, *Mutter*!"

Tragedy

In 1809 Gauss' great work, *The Theory of the Orbits of Celestial Bodies Moving in Conic Sections* came out. It describes in detail his method for finding the orbit of a planetary object from a limited number of observations. From this point on, Gauss was the most prominent astronomer of his time. That was when some scholars began referring to him as the "Prince of Mathematics."

It took forty years before the methods that Gauss explained in *The Theory of the Orbits* became the standard method for calculating orbits of planets and comets, but over time astronomers came to agree that it was the best method. Soon after its publication, Gauss was showered with honors from everywhere. His early biographer Sartorius von Waltershausen wrote that Gauss was honored by membership in scholarly societies "from the Arctic Circle to the Tropics, from the Tajo to the Ural, including the American societies."

When Gauss wrote a book, he wrote the manuscript in ink with a feather quill as he had learned from Büttner when he was eight years old, although in later years he also used a pencil. Fortunately Gauss wrote in a beautiful, clear hand, but if he needed to make a correction, it involved going back and rewriting the whole section. It was a painstaking process. Then he would give the printer his handwritten manuscript from which the printer would set the type and make a printed copy—a proof—for him to check for mistakes. *The Theory of the Orbits* was published in Latin, as was most scientific work at the time. In fact, Gauss wrote it first in German, and then he translated it into Latin because the publisher in Hamburg believed that it would sell better in Latin. As scientists later began to publish in their native tongues, Gauss lamented the trend because then scientists from all over Europe would no longer be able to read one another's works easily. It has been said that by the late 20th century, English had become the "Latin of the modern world," with scientists writing and corresponding only in English, so that once again all scientists throughout the world can communicate well. Gauss would have approved.

In 1804 the Institut de France awarded him the Lalande Medal for the best work in astronomy. Gauss accepted the medal but refused to accept any money from the French because they had conquered Braunschweig. Sophie Germain (who had first corresponded with Gauss as M. LeBlanc) intervened and bought a marvelous pendulum table clock for Gauss with part of the money. He treasured that clock for the rest of his life. For an astronomer like Gauss, an accurate clock was an indispensable scientific instrument.

In 1809 a third child, Louis, was born. He was named after the astronomer Carl Ludwig Harding, discoverer of the planetoid Juno, whose orbit Gauss also calculated. After a long and difficult labor Johanna was exhausted. Gauss was worried. His wife was not well.

"*Hännchen,* don't worry about getting things back into order here. The only thing you need to work on is your health. This has been a hard month for you. Gretel has been handling things well in the kitchen, and she says her sister Anna would be glad to come and help with the housework. I've asked her to do that so that you won't have to worry about anything. Nurse Sabina has the children under control. So all you need to do is lie there and be as beautiful as you already are without worrying about anything. Can you do that for me and the children?"

"Yes, Carl. I'll try. Could I have a sip of water? I'm so thirsty."

"Of course! I would bring you all the water in Europe if that would make you feel better. There. Do you want a little more? Or could I get you a cup of chamomile tea?"

"No, that's enough for now. You have been so wonderful about all this, but it must be hard for you to keep up with your work when you have to worry about all the details at home too. I can at least help supervise Anna."

"No, Johanna. I have only one great desire: to see you get your strength back. The only way to do that is if you free yourself of all worries. We are fine. Rest. Rest some more. Then rest some more. That is the most important thing."

"I'm worried about little Louis, Carl," said Johanna. "He doesn't nurse well—I'm afraid I'm not producing enough milk for him. Nurse Sabine says she knows a woman who could serve as a wet-nurse [a lactating mother who could give him her own breast milk]. Perhaps we should ask her to help us. Little Louis seems so weak."

"I'll ask Sabine to arrange that today," said Carl. "That will probably give you a little more strength too."

"I think it would help. Thank you, Carl," Johanna whispered.

A few days later, when Carl and Johanna were talking, Johanna was even more worried about Louis.

"Carl, Louis is sleeping more than he should," she began. "I have never seen a baby sleep so much. Even when he is awake, he seems groggy, and he isn't nursing well with the wet-nurse either. Have you noticed how strange his cry is—it seems almost frantic—and he is so pale."

"I know, Johanna," said Carl. "I'm worried too. But you need to let the rest of us worry about Louis. You need to rest."

Rest, however, was not enough. Within a month of Louis' birth, Johanna died. Gauss was devastated. Of course at that time it was well known that many women died during or soon after childbirth. But Carl never imagined that such a fate would take his beautiful wife and disrupt his happy family. "Oh, *Hännchen*! Oh, my dearest *Hännchen*!" he sighed.

He wrote to Olbers, with whom he had shared such wonderful discoveries in astronomy, to ask if he might visit him again in Bremen. "You were generous to invite me to come for a visit in Bremen when my wife is well again. Alas, she died last evening and I closed her angel eyes forever. I don't know how to deal with this. Could I come visit you now for a few weeks to try to gather my strength so that I can put together my life? All that I have to live for now is the care of my three small children."

Gauss spent the next few weeks in the welcoming home of Olbers, who cared for him both as a physician and as a friend. Although Gauss often appeared to others as glacially cold, it is obvious that Olbers felt real warmth for his young friend. As the two men talked and did astronomy together, Carl began to realize that perhaps he could continue living even though his perfect wife was gone. A tear-stained letter was later found among his papers: "Darling Johanna, we used to be so happy. You were an angel who was part of my life for five wonderful years, and now you are gone. How did I deserve you, you who were so good in every way? How shall I continue without you? Heaven help me!"

A few months later, the baby Louis also died. He had never been strong, and the last few weeks he slept almost all the time. Nevertheless, it was a shock to his already distressed father. "Poor little Louis," Gauss sighed. "Poor little Louis."

Marriage to Minna Waldeck

A year later in 1810 Gauss proposed marriage to Johanna's dear friend, Minna Waldeck. Although she was very different from Johanna, she was a good friend and a good person. She could become a suitable mother for his children, whom she already knew well. When he proposed to her, he admitted that he could offer her only a divided heart. "You must understand, *Fräulein* Waldeck, that Johanna was the love of my life. No one will ever replace her in my heart. But you are a good, kind person. If you are willing to accept a different place in my heart, I would like very much to marry you and bring up a family with you. You already know my children, and they feel comfortable in your presence." With her parents' approval, she accepted his marriage proposal.

"Carl, does your mother know about me and our marriage plans?"

"No, Minna, not yet. I'll ride to Braunschweig one day next week and tell her."

"I'd like to write her a nice letter, introducing myself and telling her about our plans. You don't mind if I do that, do you?"

"Yes, Minna. I do mind. Please don't write to my mother."

"Why not? Do you think she wouldn't approve of me? Would she worry that I am too proud or something? Would she be critical because you married again too soon after Johanna's death? Is there something wrong? What's the problem? I could write a very nice letter. I'd be glad to show it to you before I send it if that would help."

"No, Minna. I am sure you and my mother will have great admiration for each other. She will certainly approve of you."

"Then why don't you want me to write to her?"

"Because my mother couldn't read it. She doesn't know how to read. She would have to ask a neighbor or my half-brother to read it to her, and you would certainly not want to bare your beautiful soul in front of people for whom it was not intended."

"Carl, I had no idea! I didn't mean to embarrass you."

"You haven't embarrassed me, Minna. I come from a humble background. I think you already knew that. My father was a poor workingman. He never expected me to do better than succeed as a poor workingman like him. My mother is a smart and lively woman, but she is totally ignorant. However, my family brought me up to be hard-working and honest, and I am not ashamed of them."

When Gauss visited his mother in Braunschweig the following week, she was delighted to hear of his second marriage. She was pleased that he would no longer have the burden of bringing up his two small children alone, and she hoped that he and

Marriage to Minna Waldeck

Minna would have more children. She assumed that Minna was a fine woman because she knew her son could be depended on to choose another angel for his second wife. She never knew of the letter that Minna had wanted to write, and she contemplated a happy marriage for Carl and his new wife.

Carl and Minna Gauss soon had two children of their own (Eugen in 1811, and Wilhelm in 1813), and family life went on. Gauss grew very fond of his new wife Minna, and the Gauss family prospered.

However, life in the Gauss household was not always easy. Bringing up Eugen was particularly difficult. Josef and Minna had always been easy children, and Gauss and both his wives had never had to struggle with them. With Eugen, things changed.

"*Mutter*! Minna and I were playing our game very nicely, and Eugen just came over and threw all the pieces on the floor! He messed it all up. Now we can't remember where we were, and we'll have to start all over. It isn't fair! Please make him stop bothering us!"

"Eugen, come here please. Did you mess up their game?"

"Yes, *Mutter*. They wouldn't let me play."

Josef protested: "*Mutter*, we asked him if he wanted to play before we started. He's really too young for this game, but, if he wanted to, we were willing to help him. But, no! He didn't want to play our 'silly game.' So he went off to play by himself, and then the first thing we knew he came racing back in here and started throwing the pieces around."

"Eugen, we all need to respect each other. You don't want them messing up your things, do you?"

"It wouldn't bother me."

"Oh, I think it would."

"No, it wouldn't."

This was a difficult child. Minna didn't know how to handle his moods and his aggressiveness, and by the end of the day she was exhausted. The baby Wilhelm also needed lots of attention, and, in addition to all that, she was pregnant again. Josef, Minna, Eugen, and Wilhelm needed to prepare themselves for a new little brother or sister.

With four children aged ten, seven, five, and three, the mother Minna always had a crisis to deal with. Josef and little Minna, who had always been very close, were not always sure how to deal with their younger half brothers. Whenever they got started on a game, they could count on Eugen to come along and spoil it. Eugen never felt apologetic.

"*Mutter*! He's done it again! I had just been sent to London to visit the queen and Minna was waiting for her chance to climb the Alps, but now Eugen has thrown all our pieces on the floor."

"But, Josef, if you know where you were, why can't you just put your pieces back on the board and continue playing?"

"Because he'll just mess it up again! Can't you see how difficult it is?"

It was hard for Minna to deal with this situation. Eugen was her own oldest child while Josef and little Minna were her stepchildren. Of course she wanted to defend Eugen, but it was obvious that Eugen was often in the wrong. Eugen always seemed too slick, too sure of himself. He was able to manipulate her and the other children shamelessly.

"It's unfair!" shouted Josef.

Then there was Eugen's relationship with little Wilhelm. As time went on, Eugen found it increasingly easy to manipulate Wilhelm so that he would innocently compound Eugen's mischief. Wilhelm seemed sweet and docile until he suddenly turned out to be the apparent culprit. This suited Eugen just fine. He masterminded the crime, and then sat back and watched while the others blamed Wilhelm.

While his wife Minna was trying to handle the domestic situation, Gauss continued to make many new discoveries in the fields of mathematics and astronomy. Gauss was not involved in the workings of the household on a daily basis. He had time only for his work—managing the household and the children was Minna's job. Although she had help—a cook and a nurse to help with the children—she was the administrator of the family, a complex organization of sometimes difficult people. It was a full-time job.

The Trip to Munich

In 1812 Gauss published papers in a wide variety of topics, including infinite series, astronomy, optics, number theory, and the fundamental theorem of algebra. In spite of the chaos at home, he was an amazingly productive scholar. He wrote each paper with his usual meticulous attention to detail, never failing to explain perfectly the difficult concepts involved. He never went to bed before 1 A.M., and he was always up early in the morning, refining his ideas at every moment. His schedule was intense, but so was he. His piercing blue eyes were constantly identifying new concepts that needed more exploration.

In 1815, Georg IV, the Prince Regent of Hannover, awarded Gauss the Knight's Cross of the Guelph Order. Gauss prized this medal, but wore it only on occasions when he would be in the presence of the king. In 1816 King Georg granted Gauss the title *Herr Hofrat* or Royal Councillor. It demonstrated that the king respected him as one of the finest scholars in the kingdom. Gauss' colleagues often addressed him as *Hofrat*, but Gauss never used the title in referring to himself except when signing official documents.

In 1816 Gauss needed to travel to Munich to make arrangements to purchase instruments for the new observatory, which was almost complete after eight long years of construction. It was to have the most modern equipment available. Up until this time, instruments made in England were the only ones of top quality, but now some fine craftsmen were able to produce excellent instruments in Munich, relatively near to Göttingen. Gauss, having been in correspondence with the firms of Reichenbach and Utzschneider, decided to go to Munich to inspect their instruments personally even though it was only two months before the birth of Minna's third baby.

His companions for this trip were his ten-year-old son Josef and Dr. Paul Tittel, one of his students. They would be on the road for five weeks, and it looked as if it was going to be a real adventure.

Josef was delighted at the prospect of accompanying his father on this trip. Travel at this time was always slow and difficult, and the Gauss family rarely left home. For Josef to travel as "one of the men" was particularly exciting. Before they left, he had many questions, some of which his stepmother could answer.

"*Mutter*, how long will it take us to get to Munich?"

"Probably five or six days."

"Will we sleep in the carriage?"

"I suspect you will sometimes when you are traveling at night, but probably not as soundly as you sleep in your own bed."

"Will we be able to see out while we are in the carriage?"

"Oh yes, Josef. The carriage will have isinglass windows, but you will certainly see the countryside best by getting out and walking around."

"Will it be bumpy?" asked Josef.

"Yes," she answered, "but I'll provide you with blankets and pillows, so you should be pretty comfortable."

It was early spring when they set out. They had packed bags with clothing, blankets, cushions, books, and Gauss' notes on his needs for the observatory.

"Father?" asked ten-year-old Josef after they had been underway for about an hour.

"Yes, Josef?"

"I would like to see where we are going."

"Well, look out the windows. This is lovely rolling countryside," Gauss explained .

"But I can't see," protested Josef. "All I can see are the tops of the trees, and they look just like the tops of the trees in Göttingen. Could I sit up front with Olle, the driver?"

"I don't see why not, but we will have to ask Olle if he minds. Regardless, you will have to wait until we stop for a rest. We have a long distance to go, and every time we stop it takes longer," Gauss answered.

"Thank you, Olle, for letting me sit up here with you!" said Josef, sitting happily beside Olle on his bench. Then, pointing to the remains of a stone turret off to the right, Josef asked "What's that?"

"Oh, that's just some old ruin," explained Olle.

"A ruin? Of what?" asked Josef.

"Oh, some old castle or something," said Olle.

"What happened to the castle?" asked Josef. "Did it burn down?"

"Goodness, child, I don't know! It's a ruin. You'll see lots more ruins before you get to Munich. Do you see that hawk over there?" asked Olle, pointing to a large bird some distance ahead.

"Yes. Is he alive? He isn't moving at all. Can he just hang there in the air?" asked Josef.

"Yes, he's alive," Olle explained. "He's hunting. He'll hang there until he sees something that looks tasty. A hawk can see a lot better than we can. When he sees a rabbit or a mouse, he'll swoop down for the kill. The little animal on the ground will never know what hit him."

"But Olle," Josef persisted, "how does the hawk stay up there? Is he holding on to something?"

"Heck, I don't know," answered the driver. "Maybe the Professor knows something about it."

"I think I'd like to go back inside the carriage now," said Josef.

"I'm darned if I'm going to stop for that!" snapped Olle. "You wanted to ride up here, so here is where you will ride until I'm ready to let you go back inside!"

"I'm sorry," Josef said, and he was silent for a few minutes. "Olle, what are the horses' names?"

"Blessed if I know!" answered Olle. "They aren't my horses. We'll take them as far as Meiningen, and then we'll get a fresh pair for the next leg of the journey."

"Olle," asked Josef, "why is the carriage so loud? I can hardly hear you when you talk."

"Oh, that's the iron-clad wheels," Olle explained. "The metal wheels make for a rough ride, but they are safe and durable, and at least the carriage has springs. But let me tell you, child, iron wheels are a magnificent invention. They last a long time, and you have to do very little to keep them up. With wooden wheels, we would probably have to stop and replace the wheels at least once between here and Munich."

"What is the name of that town up ahead?" asked Josef, pointing down the road in front of them.

"That's Eisenach, child," said Olle. "We'll stop there for half an hour or so to give the horses some water and to let you and the Professor walk around a bit. You can move back inside the carriage then."

"That's good. I'm tired of sitting. My bottom hurts," admitted Josef.

"Yup! Mine does, too," agreed Olle.

"Father," asked Josef as his father emerged from the carriage, "how does a hawk stay up in the air?"

"That is an excellent question, Josef," Gauss answered. "Did you ask Olle?"

"Yes, and he said I should ask you," said Josef.

"I suspect it has something to do with the wind," Gauss said. "Birds are pretty light, and that makes it easier for them to fly, but most birds flap their wings when they fly. I suspect the hawk arranges it so that he is facing into the wind, and some of the wind must get trapped under his wings and that holds him up—kind of like a kite. I don't think he could do it without a strong wind. Are you going to come back inside the carriage for the next leg of the journey?"

"Yes, I think I will, but it is very interesting up there. I can really see where we are going, and I like that," said Josef.

Later that day, Josef was riding inside the carriage as they drove into the town of Bad Kissingen. "What is that funny building?" asked Josef.

"That's a spa. It's a building over a hot spring," Gauss explained. "Warm water bubbles up out of the ground here, and many people think that the spring water cures diseases. They spend a couple of weeks here every year, drinking the water, breathing in the damp air, and bathing in the water."

"Do you think the spa cures people's diseases?" asked Josef.

"I suppose it's possible," Gauss answered.

"But you don't think it's very likely, do you?" Josef asked.

"I simply don't know," admitted Gauss. "When someone is sick and can't seem to get well, I can understand that they want to do something, and going to the spa at least makes them feel as if they are doing something."

"If you were sick or if *Mutter* were sick, would you go to a spa?" asked Josef.

"Probably not, but I suppose it would depend on a lot of things," Gauss answered.

The trip took them through Würzburg on the Main River with all its lovely vineyards. Two days later in Augsburg, Gauss was fascinated by the Fuggerei—the first subsidized housing project for the poor. He had read about it, and was pleased by its utopian philosophy. Residents had to pay only a modest rent to live there, but in exchange they were required to pray daily for the souls of its founders. He explained to Josef that the doors to the Fuggerei were closed each night to protect the citizens sheltered there.

"Father, what happens to the poor people in Göttingen?" Josef asked. "Who takes care of them?"

"Most of them are able to scrape by with a meager subsistence from gathering and selling firewood or apples or something, but we don't have a real system for helping them. I expect many of them go to bed cold or hungry at night. I fear ours is a society that is not as charitable to the poor as we should be."

When they arrived in Munich, Gauss wrote a letter to his wife Minna:

> We got to Munich late yesterday, and everything has gone well. On Sunday we rode through the Thuringian Forest. It is a beautiful spot, although the roads were still icy. Then we changed horses and went as far as Meiningen where we again changed

horses for an all-night ride on the Bavarian High Road to Würzburg. We had a noonday meal in Würzburg and then rode on to Augsburg that afternoon. We spent that night in Augsburg, and then Thursday morning we did a little sightseeing in Augsburg in the morning before the last bit of the journey to Munich.

The trip has been exhausting, but we slept well last night and today we are all feeling fine. Reichenbach has already called on me, and I have spent much of today with him, looking at his sample lenses and other devices. He has urged us to stay at his home instead of the hotel, so we will pack our bags again and move there tomorrow. I believe he lives rather well just outside of town.

Gauss gave an official report of his successful trip to the university just four days before the new baby Therese's birth. His recommendations for the ordering of instruments were approved by the university, and the fine instruments were installed two and three years later in 1818 and 1819. When the craftsmen were delivering their wares, they also serviced the pendulum clock in the observatory. It had been more than a year since its last servicing, and an accurate, well-oiled clock was essential.

The New Observatory

Family life went on, with the new baby Therese fitting into the large family with three big brothers and a big sister. Fortunately she was an easy baby, and they all took turns rocking her and tending to her, but Gauss' wife Minna was not well. September of 1816 was a particularly difficult month for her.

"Carl, we have to move into the new observatory next week, don't we?" asked Minna quietly.

"Yes, Minna. We have waited eight years for this moment—the observatory is beautiful, and I think our wing will make a lovely home for us." The family wing was spacious. Its generous bedrooms, elegant living rooms, and complete kitchen made it seem more like a mansion than a mere house. They had lived well in their rental quarters, but this would be an impressive home for the distinguished professor.

"Yes, Carl, I'm sure it will be lovely. It's just that I am so tired. Little Therese always wakens me at least once during the night, Eugen seems to be harder to deal with than ever, and I just don't feel well," Minna admitted.

"You know, Minna, I've noticed that you have been coughing a lot. Do you think you are getting a cold?" asked Carl.

Observatory, Göttingen

"No, it doesn't feel like a cold. It's awfully hard for me to breathe sometimes."

"Well, we'll hire people to help with the move. You don't need to do anything about it. The maids can pack things up, and all you will have to do is tell people where to put things," said Carl.

"Even that sounds like an enormous job to me. I'm sorry, Carl. I don't mean to complain. It's just that I am so tired all the time," said Minna.

"I think I'll ask the Doctor Hartwig to look in on you. Maybe he will be able to help you," said Gauss.

"All right, Carl."

"Well, *Herr* Professor," began Dr. Hartwig, "your wife is indeed sick. I don't like her cough, and she does seem very tired. I think it may be tuberculosis. Her bright eyes are one of the telltale symptoms of that disease. I believe a trip to Bad Ems might help her. The waters there are supposed to be beneficial to people with tuberculosis."

"How long do you think we should stay?" asked Gauss.

"I would say a minimum of two weeks," Dr. Hartwig answered. "Three weeks would be better."

"All right, then. We'll do it. Thank you, Doctor."

Carl and Minna went to Bad Ems for three weeks of "cure." It was a beautiful, relaxing spot, and her doctor there was helpful. She breathed the moist air rising out of the hot springs, and she seemed to be feeling stronger.

"Carl, I think after another week here, I will be well enough to go home. Thank you so much for taking the time to bring me to the spa."

"Dr. Hartwig thought it would be helpful, and I have to admit you really do look better. You know, when Josef and I went to Munich, we passed through Bad Kissingen and Josef asked me if I thought a spa could help someone who was sick. I told him I didn't know, but I think your cure is making a difference. I am pleased to see you eating well," said Carl.

"Do you think everything will be in place in our new home in the observatory when we get back to Göttingen?" asked Minna.

"The last letter I got from Dr. Tittel indicated that the move is almost complete. I'm glad we have been able to remove you

Family Wing of the Observatory

from the scene of all that confusion. How about a walk in the park before dinner?" asked Gauss.

"Oh, I don't know. Perhaps just a little way," agreed Minna. Together they slowly walked through the park.

"Now, shall we sit on that bench for a few minutes before we walk a little farther?" asked Gauss.

"Yes, let's sit down," said Minna, "but I'm not sure that I'll be able to walk any farther today."

"Let's see after you have rested a bit. You really do need to get some exercise," said Gauss.

"Yes, Carl, that's what the doctor says, but it's so hard."

The New Observatory

When Carl and Minna returned to Göttingen, Minna was exhausted from the trip. They put her to bed in her new airy bedroom, and she rested for several days before she ventured out into the other rooms of their new home to see how things were arranged.

"Carl, it's lovely. Yes, I think we can live here very happily," said Minna.

Gauss' Mother

In 1817 Gauss' mother moved to Göttingen to live in the observatory with the Gauss family. She was 74 years old, and in the nine years since her husband's death she had been living alone in the house where Gauss had grown up. Her stepson Georg looked in on her occasionally, but they had never had a warm relationship. Before Gauss and his family moved into the observatory, the space for the family had been rather cramped, but now they had spacious living quarters. Gauss also felt that by now his mother was too old to live alone. Although he had worried about her for years, she had always been loath to leave Braunschweig. When Carl asked if she would please come and help care for his sick wife, she couldn't refuse. She had not been outside of Braunschweig since her marriage more than 40 years earlier. Although she was actually little help to her ailing daughter-in-law, Carl was pleased to have her in Göttingen.

"Carl, it is so grand here. How will I ever be comfortable? It was so nice of you to invite me to come to Göttingen to live with you, but I think I had better just go back to Braunschweig. Minna doesn't look at all well, Carl, and I'm not sure how much help I can be."

"Now, *Mutter*, we all want to have you here. I know it seems strange now, but you'll learn your way around. It's not so different from your own house. It's just a little bigger. And I'm sure your presence will make it just a little easier for Minna."

"But Carl, look at the table in the dining room. It's so big! I'll never be able to eat there. Would you mind if I eat in the kitchen with the servants? That would feel much more like home to me."

"If you want to eat in the kitchen, there is no reason that you shouldn't, but we would certainly love to have you eat with us. Maybe you should try eating in the kitchen at first, and then later on you can join us in the dining room."

She never adopted the bourgeois lifestyle of her son's family although she accepted their caring love quietly. She always ate in the kitchen. She was pleased to look out her window at the garden and the terrace of the observatory where the family had a small herb garden with dill and rosemary and parsley, and she enjoyed walking into town and visiting the markets. She particularly enjoyed buying gingerbread cookies in the market and talking with the women there.

One afternoon after she returned from a walk through the town market, she called Therese. "Therese, come here a minute, child. Look what I bought for you!"

"What do you have for me, *Oma* [Grandma]?" Therese asked her grandmother. "Oh! Wonderful! It's a gingerbread cookie! Thank you so much! Did you get it in the market?"

"Yes, they are almost as good as what I got in Braunschweig."

"*Oma*, may I go with you to the market one day? I'd like to meet the woman who makes the gingerbread."

Gauss' Mother

When Gauss' mother wanted to write a letter to her stepson Georg in Braunschweig, one of her older grandchildren was always glad to write out the letter as she dictated it.

"Josef, could you help me write a letter to Georg? It's been a long time since I have written to him."

"I'd be glad to, *Oma*. Let me go get some paper and a quill and some ink. Now I'm all set. What do you want me to write?"

"Dear Georg. Um, let's see. How are things in Braunschweig? Um... You should see how big the house in Göttingen is. I never thought I would live so grandly... The children are lovely, and I enjoy watching Therese play. She is a dear child."

"Thank you for your help, Josef. May I look at it? Oh, it looks lovely. Georg should be pleased to get it. You are a fine boy to help me so nicely."

Many afternoons, Gauss, his mother, and Minna had coffee in Minna's room with Minna propped on pillows in her bed. One afternoon, five-year-old Therese (who usually joined them as well) was standing at the window, watching the rain fall. The normally happy child chanted,

> "*Regen, regen, geh sofort!*
> *Geh an einen anderen Ort!*"
>
> [Rain, rain, go away at once!
> Go to some other place!]

"Come here, little one," said her father. "Do you know what an American child would say to an all-day rain like this?"

"What?"

"Rain, rain, go away!
Come again some other day!"

"Does it mean the same thing?"

"Almost. The first line is exactly the same, but in the second line the American child tells the rain to come back another day while you have told it to go to another town instead."

"I don't want it to come back tomorrow. I like my version better. I don't want it ever to rain again in Göttingen!"

"But, Therese, we can't get along without rain. How would the flowers grow? How would you take a bath? Where would we find water for your mother to drink?"

"Okay. I guess it can rain sometimes. But not when I want to play outside. Could I have another piece of cake?"

"What did you say, Therese?" asked her mother.

"Please?"

"Of course you may. It is very good, isn't it?"

Gauss loved these late afternoon family times, and he found little Therese a constant joy. It was hard with Minna so sick, but little Therese always provided some charming diversion and he knew that his mother, who was rapidly losing her sight, was comfortable and content.

"*Oma*," said Therese, "I like it when you tell me a story. Would you tell me the story of Little Red Riding Hood again?"

"Yes, come over and sit next to me. Once upon a time, and it was a very rainy time, Little Red Riding Hood was heading out into the woods with her basket full of gingerbread for her grandmother."

"*Oma!* Does it have to rain in the story too? Won't Little Red Riding Hood get wet?"

"Don't forget, Therese, that she has on her red cape. That would keep her warm and dry. And she was taking gingerbread to her grandmother—lovely big fat ones."

Part IV

Surveyor of Hannover, Father of a Growing Family
1815–1832

Surveying

In 1816 Gauss became involved in a major mapping project. One of his first students, Heinrich Christian Schumacher, was a professor of astronomy at the university in Copenhagen and he had begun a detailed survey of the kingdom of Denmark so that he could make an accurate map. He wrote to Gauss, asking for some improvements on the general approach to surveying and inviting him to extend this project to the south. Gauss was intrigued. He had tried out some geodesy when he was still living in Braunschweig, and in 1799 the Prussian Colonel Lecoq had consulted with Gauss when he needed some help with trigonometry on his early attempts at a survey of Braunschweig. Between 1800 and 1805 Gauss had also done some fieldwork with Zach's sextant, experimenting with taking measurements.

Gauss wrote back to Schumacher, detailing some new techniques that he had devised and saying that he thought the mapping project would yield beautiful results, but he was concerned about how to arrange the funding for such a project. He also worried about how to find good, reliable helpers.

Gauss eventually concluded that military officers were the best assistants for this kind of surveying because the peasants who

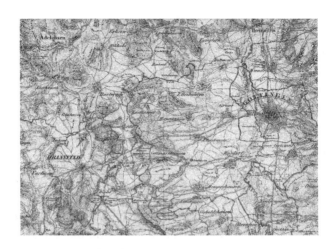

Papen's Atlas, map of Göttingen and environs

would be working with them would have automatic respect for the officers. Schumacher wrote directly to the Prince in Hannover to establish Gauss' official part of this project.

In 1818 Gauss announced that he had been formally requested to do a survey of Lüneburg. After buying a carriage, he met Schumacher and an assistant at Lüneburg, and together they measured angles from the tower of St. Michael's Church at Lüneburg to Hamburg and Lauenburg. The measuring went fairly well, but there was a problem with the sightings because of an annoying reflection of the sun on the windows of the church in Hamburg.

"Yes, I see the church now. Oh, drat! The reflection of the sun in the window is blinding! Yes. I've got a good reading here. Oh! If only it weren't so bright!" complained Schumacher.

Surveying

"Yes, Schumacher, the glare makes it difficult," Gauss said. "But, wait. The window is able to shoot that beam of light straight at us. Do you suppose we could shoot a beam of light straight at the church with the sun at a different angle? Maybe if a colleague were at the church window in Hamburg, he could beam a light at us here in Lüneburg and measure the angle? No, that's too complicated. I believe we could do it all from here. If we had a mirror, couldn't we reflect the light of the sun to hit the church in Hamburg? The mirror would not need to be too big. I'll bet it's only one pane of glass that is making this annoying glare today. The mirror could probably be so small that it would fit in the palm of my hand. Then we would need to be able to see exactly where the beam of light hits, so I guess we would need a telescope of sorts. If we could then measure the angle at which the beam has to go to hit the target, we would have all that we need. I think I've got something here."

"*Herr* Professor, do you really think this is practical?" asked Schumacher.

"I'm sure of it. I'll get my mechanic, Rumpf, to make me a prototype."

"*Herr* Professor, is his name really Rumpf? You didn't make that up, did you?"

"No, Schumacher, I didn't make it up. His name really is Rumpf."

"Is he a trunk, built thick and rough like a tree trunk?"

"Schumacher, I think you are being a little hard on him. He is a simple man whose tools are his friends. I have rarely seen a man so skillful. You'll see that my new device will come out just the way I draw it. Rumpf is a real gem. I believe I'll call this device a heliotrope because it will follow the sun just as the flower does."

In the next few years Gauss needed to make only a few improvements on his heliotrope, and it helped him greatly in his surveying. It became the standard tool for surveying within only a few years, and it wasn't replaced until the development of satellite photography and the use of laser beams (which allow the measuring of distances as well as angles). The heliotropes constructed by Rumpf were just as good as Gauss had predicted. He soon contracted with a firm in Hamburg to make several more heliotropes. Surprisingly, the mirror itself was only the size of a modern day credit card. Gauss declared that if he could make a mirror that was big enough, he could use it to beam a ray onto the moon itself! Rumpf's first heliotrope was ready in July 1821.

In 1820, King Georg IV of England and Hannover provided a grant to pay for the survey, and Gauss spent a large part of the next ten years in the field. Gauss' 14-year-old son Josef began working with him in the summer of 1822, and later Josef took over some of the fieldwork himself. He liked the work so well that he decided to abandon his plan to study law at Göttingen and instead to go to the artillery school at Hannover. He enjoyed working with the soldiers on the project, and he had a natural talent for technical work. Gauss approved of this plan, because he saw that Josef was not a born scholar.

Surveying at the time was based on visible measurements using trigonometric points and triangulation. The surveyors attempted to verify each point from at least two different direc-

Lüneburg Heath: terrain where surveying is extremely difficult because of dense vegetation with no hills

tions. In addition, they established triangulation on a larger scale to keep the smaller triangulation in line. Where the land was hilly or mostly open, this was relatively easy to arrange, but a flat wooded area was difficult. The Lüneburg Heath, with its lovely lavender heather, north of Hannover and Braunschweig, had been completely omitted in the Napoleonic surveys because of the difficulties involved, but with the combination of the heliotrope and his determination and skill, Gauss succeeded.

Gauss decided to use the observatory in Göttingen as the origination point for his triangulation. A brass marker that is now hidden under a trap door in the floor of the observatory shows the exact point. That point is one vertex of the first triangles that Gauss drew. He proudly showed it to his mother: "*Mutter*, this point is the anchor from which I will make all my measurements. The actual point is the very center of this brass

Brass marker: origination point for Gauss' survey of northern Germany

circle. Although all my measurements are abstract, this point is substantial."

His mother was impressed. "Carl, you have done so many important things! I can't tell you how proud I am!"

A hill at Weende just five kilometers north of Göttingen had a good view of Göttingen and provided the first sightings. Gauss traveled to the town of Dransfeld, 15 kilometers west of Göttingen and walked up the Hohehagen hill nearby. The French had built a signal tower which had since fallen down on top of Hohehagen, but, because it provided such a good view, Gauss was determined to have a new tower built there to provide good sightings. He had another tower built on the hill called Hils, near Hannover. Using these three points—Weende, Hils, and Hohehagen, each visible from the other two—Gauss set up his beginning triangle.

Gauss' velvet cap

In July 1819 Gauss was forced to break off his surveying because of heat exhaustion.

"I'm sorry, men. I can't take this heat any longer."

"But *Herr* Professor, perhaps if you dressed a little less formally you would be more comfortable."

"No. When I am out working, I wear my wool vest and coat, and quite naturally I wear my velvet cap. A professor and scholar is not seen outside his home in casual attire."

"But perhaps you could loosen your collar, *Herr* Professor."

"That will not be necessary."

"But *Herr* Professor, think of how you are suffering."

"That is precisely why I propose to go home and rest. Sometimes you show remarkably little intelligence." Gauss was not always easy to get along with.

When he arrived at home, his mother, who was completely blind by this time, was upset. "Carl, are you all right?" she asked. She never understood why he had to be away from home for extended periods. She didn't like having a change in the family routine, and when he was gone she felt lost. She loved his wife and the children, but her son was the focus of her life.

Summers on the Road

For seven or eight summers, Gauss spent most of his time on the road, surveying. He and his team would take measurements all day, forcing their way through dense vegetation as they tried to get the most precise measurements possible.

One unpleasant challenge was arranging with peasants the cost of removing a tree or two that might be blocking the view from one triangulation point to the next.

"Sir," Gauss began, "we need to cut down these two trees in order to get precise measurements for our triangulation project, which will lead to an accurate map of the area."

"Now, wait a minute, young man. Just exactly who are you and why do you want to cut down my fine oak trees?" Gauss' assistants bristled at that term of address. After all, this ignorant man was addressing the professor himself! Furthermore, he was being rude. However, Gauss continued to address him politely and patiently.

"I am Professor Doctor Carl Friedrich Gauss, head of the observatory at the university in Göttingen. Here is the order from the king, authorizing this project. He wishes to have a precise map of the entire Kingdom of Hannover, and in order to do that we have to make measurements from some distance. When the woods are thick, it is particularly difficult. Napoleon was unable to make this map, but German ingenuity will succeed where the French failed."

"But those are my oaks you're talking about! Do you know how long it takes to grow a tree like that, young man?"

"Yes, sir. We are prepared to compensate you for the loss of these two trees. Would you consider five *Thalers* a reasonable amount?"

"Five *Thalers* per tree?"

"All right, then. Five *Thalers* per tree. That should more than make up for any profit you would have made on the trees in the future."

"Oh, all right. Get it done fast, mind you, and stack the wood neatly. Don't leave a mess for me to clean up later!"

Every measurement had to be made several times. In the evening, Gauss applied his method of least squares to the data, using probability to determine the best estimate of the actual measurement. Gauss calculated with lightning speed, but still the calculations were a trial. Papen's Atlas, which was first issued in 1832, was based on 3,000 coordinates, each one of which was the result of Gauss' taking many figures and distilling them into the one most likely measurement for that particular sighting.

Gauss often found an inn at which to stay near the points of his triangulation. In Dransfeld, Gauss stayed at the Hotel zur Krone, a pleasant inn with a good kitchen. His room was comfortable, the candleholders were adequate and plentiful, and the table in his room was reasonably comfortable for his late night work. When he arose in the morning, he found a hearty breakfast served in a homey atmosphere. However, this was not always the case.

On the Lüneburg Heath, sometimes he boarded with a peasant family and he was lucky if he had a bed to himself. Gauss did not enjoy sharing a bed with the farmer's younger son. He sometimes had trouble finding a basin in which to wash his face in the morning, and he often had to cope with a chamber pot that had not been emptied at the appropriate time. Waiting to begin his evening's calculations at the dining table until the evening meal had been cleaned up was another frustrating delay. If he brought his own candles, they might disappear from one evening to the next, and the mounting of the candles was often a serious challenge. Setting his hosts' house on fire would not have helped the project!

He did not enjoy his time away from home. It was an exhausting routine, and he often complained of the conditions, particularly the heat. His companions described him on several occasions with sweat pouring down his brow but velvet cap firmly in place!

Gauss' work on geodesy was closely related to his work on geometry—specifically the process of portraying a curved surface onto a flat surface. In 1828 Gauss was able to define the surface of the earth as the curved surface where every point is perpendicular to the direction of gravity. Of course the surface of the earth appears to us to be a plane until we consider it on a large scale and see that it is in fact close to a sphere. This is why the triangulations both on a small scale and on a large scale were crucial.

In 1823, while Gauss was on the road with his surveying, he was thrown from a horse that had not been sufficiently broken in.

"*Herr* Professor! Are you all right? That was a nasty fall!"

"Yes, I think I'm all right. What is the matter with that horse?"

"I'm sorry, *Herr* Professor. The horse is definitely frisky. No, sir, I don't think you should get up just yet. Let me use my clean handkerchief for that cut on your nose. I fear that you will have a colorful black eye from this."

"I'm all right. However, I think I will return to my lodgings for the rest of the morning. Can you carry on with the measurements?"

"Yes, *Herr* Professor. Josef, please conduct your father back to the inn."

Gauss was shaken by the experience. The resultant black eye and the cuts and bruises on his arm and nose healed quickly, but Gauss didn't like it. He was convinced that it was safest to stay at home, quietly doing the mathematics for which he was best suited.

In July, he was suddenly called home because his wife, Minna, was gravely ill. Although he cut short his current measurements, he couldn't resist stopping on the way home to measure positions of 100 church towers and other spots in the Hildesheim area. Getting home to Minna was important, but he couldn't simply ride past those important locations without stopping! She could wait just a little longer for his return.

"Carl, I'm so glad you've come home. Sometimes I don't think I can cope any longer. I'm so tired—so tired all the time."

"I'm sorry it took so long for me to get here, Minna. I had to stop on the way and take a few more measurements, but I came as quickly as I could. You know, your eyes are so bright!"

"Do my eyes look bright? They don't feel bright. None of me feels bright at all. I'm so tired."

"I wonder if tea is ready. Let me go to the kitchen and see. I'll ask Mother if she would like to join us for tea."

Just then Gretchen bustled in with the tea tray. A few minutes later, Therese led her blind grandmother in to join them. "*Herr* Professor, *Frau* Professor, *Frau* Gauss, and *Fräulein* Gauss [Therese], tea is ready. Here is the lemon and sugar, and I made your favorite chocolate cake, *Frau* Professor. Here is your tea, *Frau* Gauss. Can I give you a piece of cake too?"

"Thank you, Gretchen," Gauss' mother replied.

"Professor? Can I serve you some tea and a piece of cake?" Gretchen asked.

"Yes, Gretchen. That would be wonderful," Gauss answered.

"*Frau* Professor? Can I serve you some cake?" she asked Gauss' wife.

"No thank you, Gretchen, I'm not hungry. Just a little tea, please."

"But Minna, this cake is delicious! *Mutter*, don't you find it delicious? Won't you try just a bit, Minna?" asked Gauss.

"No, I just don't feel like eating. I'm sorry."

"*Frau* Professor, don't you remember that the doctor said you need to be eating more? It will make you stronger," Gretchen urged.

Just then Minna began to cough. Her coughs were deep and wrenching. Gretchen provided a handkerchief for her to cough into, and when the spasm had passed, there was a great deal of blood on the handkerchief. Gauss' wife was indeed very sick.

"Minna, perhaps another sip of tea would help."

"I'll try just a bit. Yes, that's good. Thank you."

"And how about just one bite of this wonderful cake? It's your favorite, I know."

"No, Carl. I can't. I'm sorry."

"But Minna, you are so thin. Can't you eat just a little bit? I think you are starving yourself."

"One bite. That's all I can take. Yes, it tastes good. You are right."

"Then how about one more little bite?"

"Thank you. That's all. I'm so tired. I think I'll close my eyes for a few minutes now."

In 1823 Gauss won the prize of the Royal Society of Copenhagen. Because of Minna's illness and the need to hire help for her and care for the children and his mother, Gauss requested that the medal be converted to cash so that he could pay his bills.

In 1824 Gauss was again called home, this time because his younger children all had the measles, and so did his already sick wife. Fortunately his mother was spared.

At first, only Eugen was sick. He was miserable, and he made sure everyone else was miserable too. "No, I don't want any lunch. No, don't touch me. Just leave me alone. The sun is so bright! It hurts my eyes. Go away! I'm cold. Bring me some tea. Ugh! It tastes awful! I'm all stuffed up. I can't breathe! Leave me alone! Oh, I itch. My tummy itches all over! I'm so hot! Isn't there anything in this house that is cool? Go away! Don't touch me. Bring me something cool to drink."

At first they thought he just had a cold, but then he developed a fever and his eyes became red and swollen. Soon he was covered with a rash. Within a week, he began to feel better, his fever subsided, and he resumed his usual childhood pranks. Just as he was recovering, little Wilhelm and Therese began to cough and sneeze along with their already ailing mother. While Eugen romped and whooped, the others languished in bed, utterly miserable. Minna's condition deteriorated to the point that the doctor feared she would die. Gauss handled the crisis as well as he could, and they all survived.

At this time, Berlin tried once more to lure Gauss to work at the institute there. However, the university at Göttingen was not prepared to lose its most famous scholar, and so they increased Gauss' salary to 2,500 *Thalers*—a generous salary for a professor. The new salary was enough to convince Gauss to stay in Göttingen. While he didn't want to move anyway, he also needed to protect his sick wife and his blind mother from the inconvenience of moving.

In early 1825 he bought a beautiful new carriage in Hannover. However, in April of that year a wheel got caught in a rut and the carriage turned over. A box with a new theodolite (a precision instrument for measuring angles of inclination) hit Gauss in the thigh, causing him genuine pain but no lasting injuries. Fortunately, the theodolite was not damaged.

"Yes, yes, I'm all right. Oh! My leg! Is there something wrong with the construction of this carriage?"

"No, *Herr* Professor," the driver replied. "The wheel got caught in a rut. I think we can get the carriage repaired. It doesn't seem to be badly damaged. We should be able to get you on your way as soon as help arrives."

"On my way? No, sir! I am going home," Gauss announced.

"Can you put any weight on that leg, sir?"

"Yes, I guess I can. I guess it is not broken after all."

Although Gauss' health was always good, he was something of a hypochondriac. He worried about little aches and pains, his reactions to the heat, and his occasional bouts with constipation. He was always concerned that these minor ailments would turn into something major. He did not go so far as to consult a doctor except for his occasional confidences to Olbers, his astronomer/doctor friend in Bremen. He just worried. In comparison to Minna's illness, his complaints were trivial.

Trouble with Eugen

As Gauss' second son Eugen grew up, he continued to vex his parents. It was clear to everyone that Eugen was brilliant—clearly a gifted child—and Gauss decided early on that he should study law, but Eugen had other ideas.

One evening Gauss asked his second son to come talk with him in his study. "Eugen, I see that you are doing well in school. Your teacher seems to think that you have real promise."

"I guess I'm doing all right, Father," Eugen admitted.

"Eugen, I'd like to see you study law at the university. I think you could make a fine career as a lawyer. Perhaps you could follow in your grandfather Waldeck's footsteps and eventually teach on the law faculty here at Göttingen. It would make me happy to see you set out on such a path."

"To study law?" Eugen asked in horror. "Why would I want to do that? I have no interest in picky little legal arguments. I don't want to go to court to argue about stupid little disputes. Mathematics is what interests me, Father. I would like to study mathematics."

"No, no, my son. That is not a good plan. Very few people are able to make careers as mathematicians. It is not at all practical. There are so few jobs. You have so many excellent talents.

You must follow a career that will lead to a good and profitable life. Law is an excellent field, and you would be good at it."

"What would you have said if your father had insisted that you become a lawyer? Would you have done it?"

"My father knew nothing about such things. He would not have had an opinion on this," Gauss answered.

"No, but if he had, how would you have felt?"

"I believe I would have listened to him," said Gauss, "and I believe you should listen to me. I think we have gone far enough in this discussion. I would like you to enter the university next fall and study law." As Eugen left the room, Gauss thought back to his relation with his own father. No, he had to admit to himself, he had not followed his father's wishes, but his father was ignorant. His father would have had no understanding of the choice between a career in law and one in mathematics. He would have approved of neither of them.

Gauss and his wife Minna had no idea how to handle Eugen. Gauss was afraid that his son would not measure up as a mathematician. Gauss set a high standard, and anyone with the name Gauss who was a mathematician must be the very best. If young Eugen was so bright, he ought to prepare himself for a real career.

"Carl, would Eugen listen to you?" asked his fearful mother when Gauss went in to talk with her in her bedroom.

"No. I don't know what to do with the boy," said Gauss. "He is as stubborn as a mule. He wants only one thing, and I know that he is wrong. He should not be a mathematician. I am sure he is a very gifted young man, but law is the best career for him. However, he refuses to hear me."

"Carl, maybe you should let him study mathematics," suggested Minna. "He does seem very talented. Not everyone is meant for a career in law. What makes you so sure he can't do mathematics?"

"Now you're starting to sound like Eugen!" Gauss snapped.

"But he wants it so much. Perhaps he has inherited more talent than you suspect. How will we ever know unless he tries? I know that you know what it is to be a mathematician, and I really don't. However, I just wonder if he couldn't do it."

"No, Minna. It is a bad plan. He is not cut out for a life in mathematics. He is too impetuous, too reckless, too sloppy. No. He could never be a first-class mathematician. I won't hear of it. You must accept that," announced Gauss.

Eugen entered the university according to his father's wishes, but immediately took up a life of gambling and drinking instead of law, and, despite his parents' earnest wishes for his success in law, got himself deeper and deeper in trouble. When he finally realized in 1830 that he was irretrievably deep in debt, he presented his father with a statement of all his debts and announced that he was leaving for America.

His father was livid. His mother was distraught. His blind grandmother couldn't understand what was happening.

For many years, Gauss' wife Minna had suffered from what were called "nerves" in addition to tuberculosis. She was miserable. She was lonely. She was depressed. Gauss wanted to help her, but interacting with his family was not his strength, and he was, as always, possessed by his work. He was kind in his own way, but there were limits. Minna was unhappy and she had no reserves of strength. Thus she wallowed in her unhappiness and became more and more depressed. She would apologize for her unhappiness, and then she would be all the more unhappy, the more she tried to explain. Gauss' mother didn't know what to think.

"Minna, dear, perhaps a cup of tea would make you feel better," her mother-in-law suggested.

"No, not now. Thank you, *Frau* Gauss," Minna answered. Minna, an articulate, educated woman, found it difficult to deal with her ignorant mother-in-law. Her simple bits of homey wisdom seemed inane to the gravely ill woman.

At this time, Gauss' survey of northern Germany was nearing an end, and he was staying at home in Göttingen while his oldest son Josef supervised the actual surveying. With Minna very sick, Gauss was tired of being away. He had always had difficulty making arrangements for his wife and his blind mother. His daughters Minna and Therese were responsible for much of his wife's daily care, but he also hired other women to help out. Young Minna and Therese certainly deserved some relief in the care of their ailing mother and grandmother.

One evening in 1831, Gauss was in his room processing numbers from the survey. A servant came into Gauss' study and said, "*Herr* Professor, *Frau* Professor has asked me to call you. She is very weak. She would like you to go to her at once."

"Tell her to wait a minute. I've got to finish this calculation."

Trouble with Eugen

It was typical of Gauss that he was so completely involved in his mathematical work that he was truly oblivious to everything else. It was not that he didn't care about his wife's suffering. He loved her deeply. It was just that he worked with total concentration, and he could concentrate totally on only one thing at a time.

The servant rang a second time. "*Herr* Professor, *Frau* Professor needs you urgently. I fear she is dying."

"What? Oh, yes. Minna needs me. All right. I will go to her. Thank you."

After being bed-ridden for several years, his wife Minna finally died of tuberculosis in 1831. It was a lingering death, affecting the family profoundly. Gauss' mother sat with her day after day, providing a sympathetic, human presence in the death watch. Her granddaughters and the hired nurses often had to work around the older woman as they cared for Minna. Gauss wrote about Minna's death to Schumacher: "A week ago I buried my wife. Her body had caused her so much pain! Now she has given it back to the earth, and finally her suffering is over."

Her daughters Minna and Therese suddenly found themselves set free. They had devoted years to their mother's care, and they hardly knew what to do with themselves now. It took a few months for them to begin to focus on their own lives once again.

Several months before his mother's death, Eugen wrote a letter to his father. He described the sea journey, reported on his life as a soldier, and asked if his father could please send him some money so that he could buy a commission as an officer. The response was a resounding no! Gauss informed his second son that his mother was close to death, and that Eugen could expect nothing from him until he was able to prove that he had become a hard-working, moral young man. Gauss later wrote him to tell him of his mother's death, and Eugen sent condolences and thanks to the family for caring for his mother.

Over time, Eugen was able to provide letters of approval from his superiors, and the letters between father and son grew friendlier. A few years later, Eugen married, and soon he and his wife were well established with a large family. Eugen was a millionaire by the time he died. Gauss began to look forward to hearing from his once dissolute son in America.

By the time Eugen was 80 years old, long after his father's death, Eugen amused himself by calculating in his head the value of $1 if it had been invested at the time of Adam and Eve at a rate of 4% interest per year. Since he was blind by that time, he asked his son to write down the current total every few days because he didn't dare trust his memory. In fact this wasn't necessary. At least once, Eugen had to correct his son's figures. When he finished the calculation, the sum was equal in value to a gold cube bigger than the known universe. Eugen's father, Carl Friedrich Gauss, may have underestimated his son's mathematical abilities and powers of concentration.

In 1837 Gauss' youngest son Wilhelm also emigrated to America with his bride, the niece of an astronomer whom Gauss met through Olbers. Although Wilhelm's departure was not accompanied by the rancor associated with Eugen's departure, it was still difficult for Gauss. He was disappointed to lose regular contact with Wilhelm, with whom he had worked closely for several years. Since Gauss never saw either of his two American sons again, he never met his numerous grandchildren in America. Wilhelm too was a very wealthy man by the time he died.

Part V

Magnetic Professor, Prince of Mathematics 1833–1855

Non-Euclidean Geometry

Gauss had written in his diary in September 1799, "In the principles of geometry we have made outstanding progress." He continued to be fascinated by geometry throughout his life.

Bolyai, Gauss' friend from his student days, made several attempts at proving the parallel postulate in the first ten years after he left Göttingen. When he sent his first serious attempt to Gauss in 1804, Gauss returned it to him with some suggestions and the comment that he planned to settle the question before he died.

Although Bolyai sent him another attempt at proving the postulate in 1808, Gauss didn't respond. When Bolyai's son Johann announced that he planned to tackle it, Bolyai urged him not to waste his time on it. In 1825 Johann Bolyai wrote a full explanation of his view of non-Euclidean geometry and his father proudly attached it to a book he was having published at that time. His father sent a copy of the book to Gauss, and Bolyai and his son waited impatiently for Gauss to publicly announce Johann's brilliant discovery of non-Euclidean geometry.

Instead, Gauss responded in a personal letter: "To praise Johann's fine work would be only to praise myself since I discovered non-Euclidean geometry more than 20 years ago!" Young

Johann Bolyai was deeply offended. This caused a lasting rift between father and son although not between Gauss and Bolyai, who would not hear anything critical of his lifelong friend— the world's greatest living mathematician. Wolfgang Bolyai took Gauss at his word. If Gauss said that he had discovered non-Euclidean geometry many years before, Bolyai knew it was true. He knew that Gauss was an honest scholar.

Gauss also stayed in close contact with his student Schumacher even after they had completed the measurements for the map of Northern Germany and Denmark. They had a healthy respect for each other, and both men enjoyed Schumacher's occasional visits to Göttingen.

"Schumacher, are you familiar with the work of Lobachevsky?" Gauss asked his colleague and former student on one of his visits to Göttingen in 1839.

"Yes. He's Russian, and doing very interesting work on non-Euclidean geometry. Does that interest you, *Herr* Professor?"

"As a matter of fact, it does."

"You haven't done any work in that field, have you?" asked Schumacher.

"Oh, yes. In 1795 I spent some considerable time on geometry. I concluded that it was not possible to prove the parallel postulate and that it was therefore necessary to entertain the concept of a geometry where it is possible to draw more than one line parallel to a given line through a given point. I realized that there also had to be triangles whose three angles add up to more or less than 180° in the non-Euclidean world. I had it all mapped out."

"But you never published anything on non-Euclidean geometry. Why not?" continued Schumacher.

"Well, as you know, I don't like to rock the boat. Euclid was still held in such ridiculous awe that I figured any attempt to suggest an alternative geometry would only cause a furor, so I kept quiet. I knew non-Euclidean geometry was true, and my construct was perfectly consistent with Euclid's first four postulates."

"This is amazing, *Herr* Professor! You have known about it for more than 30 years and you have never written about it or talked about it!"

"It didn't seem terribly important. I knew it was true and then I moved on to other things."

Gauss was a man who thought carefully and precisely, and once he was convinced of the truth of his views he was satisfied. He didn't need some other authoritative figure to corroborate his views. His confidence in his own mathematics was unshakeable. When he was ten years old, there was no question in his mind that he had correctly found the sum of the first 100 numbers despite the suspicious stares of *Herr* Büttner. In fact, his mathematical conclusions have all stood the test of time. If Gauss proved it, it was true.

"But what I was getting at, Schumacher," Gauss continued, "is that I would like to see what Lobachevsky has done. By the way, did I tell you that I have been flirting with learning Sanskrit?"

"No. I understand it is a difficult language. It's related to Finnish, isn't it?" asked Schumacher.

"I believe so. Well, I'm not sure that Sanskrit is more difficult than other languages, but I must say I don't find it particularly engaging. I believe I will give up on Sanskrit and take up the serious study of Russian. Then I could read Lobachevsky in the original. I dislike depending on the translations of others."

In the following year, Gauss learned to read, write, and speak Russian so that he could read Lobachevsky. At that time, he was the only resident of Göttingen who knew Russian, and his Russian was flawless. He had not lost his gift for languages.

Gauss continued, "Did I tell you, Schumacher, that Lobachevsky is a student of Martin Bartels, my childhood tutor and friend?" Gauss and Bartels had not seen each other since the year that Gauss' benefactor, the Duke, had died. However, they had kept in touch through letters. Gauss naturally had some personal interest in Lobachevsky whose "Doctor-Father" was his well-respected childhood friend and tutor. That made Gauss an intellectual uncle to Lobachevsky.

"*Herr* Professor, I'd like to ask you about something else," said Schumacher. "How do you calculate so rapidly? When I first met you, I saw at once that you were an outstanding scholar, and certainly others had said that about you too. But when I saw you in action, I was amazed that you were also an amazing calculator. Do you have a special technique?"

"I never sat down and figured out a specific plan for calculating," Gauss replied, "but certain relationships among numbers have always stuck in my mind. I probably know a lot more products, like $13 \times 29 = 377$ or $19 \times 53 = 1007$ than most people know. Those tidbits allow me to move faster than others could, but I have to admit that I sometimes get tired of doing long calculations."

Magnetism

In 1804 Alexander von Humboldt, the world-renowned naturalist and explorer, discovered that wherever he went in America, scholars asked him for information about Gauss. Humboldt, astonished that American scholars knew far more about his neighbor than he did, determined that he would make Gauss' acquaintance.

When Humboldt wrote to Gauss in 1807, it was the beginning of a warm friendship. When the king of Prussia invited Humboldt to enter the Berlin Academy of Sciences so that it would have the prestige associated with his name, Humboldt replied that the only name that could do that was Gauss'. In 1828, he invited Gauss to be a guest at his home during an international congress in Berlin for three weeks. Humboldt had been trying to lure Gauss to move to Berlin for some time, but those three weeks were the most that Humboldt would get. The visit was a great pleasure to the two men, both of whom were highly esteemed scholars. It was at this time that Gauss first met the young physicist Wilhelm Weber.

In 1830 the chairman of the physics department at Göttingen University, Johann Tobias Mayer, died and his position became available. Gauss was asked for his recommendation for a new man for the post. Gauss responded that the physicist needed to be a thoroughly competent physicist but also an excellent mathematician. He needed to be able to both teach and administer. Gauss ended up recommending a total of four people, whom he described in some detail. At the end of his confidential memo, he wrote this summary:

> If we want a 60-year-old, well-known, established scholar and we are willing to pay for him, I recommend Bohnenberger. If we want a pure administrator who knows how to work with people and get things done, I recommend Gerling. If we want to hire the only one of these men who has actually applied for the job, I recommend Seeber. If we want a genius who will probably do original work that will continue Göttingen's tradition of excellence in research, I recommend Weber. He is 27 years old, and is just beginning what promises to be a brilliant career. He is also an excellent lecturer.

There was no doubt which candidate Gauss favored. Weber was offered the position in 1831, and it was the beginning of a warm and fruitful scientific and personal association. Weber was an experimental scientist while Gauss was more theoretical, and the two happily approached problems together from their different perspectives. At this time, when Gauss' wife, Minna, was on her deathbed, his home was not a particularly congenial place to be, so he often had dinner at Weber's home or invited Weber to have dinner with him at his home. Weber's sister Lina, who kept house for Weber, sometimes resented Gauss' monopolization of her brother's life. "Professor Gauss works with Wilhelm all day, but after a full day of work, he often comes here for dinner. He is a formal man, who believes that he should not discuss his current research at the dinner table, and he often stays for hours after dinner, making small talk and being polite. Last week he came for dinner three days in a row."

Although Gauss had expressed an interest in the study of magnetism as early as 1803 in a letter to Olbers, and he had been leaning in that direction beginning in 1804 after reading Humboldt's popular descriptions of his visit to America, it wasn't until Weber's arrival in Göttingen that Gauss seriously studied magnetism. In January 1833 Gauss wrote a paper on magnetic forces within the earth, and in it he established the three fundamental measurements of magnetism: mass, length, and time. For many years, the standard unit of magnetic force was the gauss; it has now been replaced by the metric unit the tesla, which is equal to 10,000 gauss. Gauss discovered that magnetic force is also influenced by temperature.

In 1833 Gauss sent an official memo to the university board proposing the construction of a magnetic laboratory at Göttingen. The board immediately approved its construction, and the laboratory was constructed on the grounds of the observatory according to specifications written by Gauss and Weber. In the building, everything that normally would have been made of iron or steel was made instead of copper. The ceilings were ten feet high, and all doors and windows were double with an inner and an outer layer in order to minimize air currents. The

Statue of Gauss and Weber, Göttingen

laboratory was very expensive to build, primarily because of the costly substitution of copper for iron and steel. It was ready for use in the fall of that year. At this time, Gauss' youngest son, Wilhelm, who was then 20 years old and had not yet left for America, helped with observations in the lab. Measurements of magnetic force had to be recorded at 8 A.M. and 1 P.M. because these represented the two extremes of magnetic force during the day.

The magnetic lab became an instant sensation among scientific circles, and it was not long before there were similar labs all over Europe. The magnetic lab at the Royal Observatory in Greenwich, England, was completed in 1838. Gauss and Weber established a magnetic association for the propagation of information on magnetism. Gauss and Weber wrote more than two-thirds of the articles in the association's journal, and it had wide influence. In 1841 an American sea captain, Charles Wilkes, found the magnetic South Pole almost exactly where Gauss had theorized that it must be, and the English Captain James Clark Ross located the magnetic North Pole in 1831, similarly close to Gauss' calculated location. In fact, Gauss had done the preliminary calculations necessary for the theory of terrestrial magnetism in 1806, but it wasn't until 30 years later that he and Weber had the experimental data necessary to prove it. Gauss found that most of earth's magnetism comes from deep inside the earth, and that the earth itself is far less magnetic than an ordinary piece of steel.

During World War II, when the Nazis made mines that were set off by the magnetic field of a ship, the Allies devised a process for demagnetizing the ships that was called *degaussing* after Carl Friedrich Gauss. Computers in the year 2000 had a special program called degaussing to reduce the magnetism in the computer's monitor. In name as well as theory, Gauss had a lasting influence on the whole field of magnetism, and astrophysicists continue to reckon cosmic magnetic fields in units of gauss.

In the 1830s Gauss continued his friendship with Olbers, who was always interested in hearing about Gauss' newest discoveries. Electricity, closely connected with magnetism, was one of those topics. In 1837, Gauss and Olbers were amazed to read of the remarkable uses electricity was finding in America. They were skeptical but fascinated. Could electricity actually be used to power a printing press? Could it play a role in medicine? They decided that it was probably mostly exaggeration, but perhaps there was some truth in it.

Gauss and Weber also collaborated on the development of the world's first working system of telegraphy. After Gauss discovered that he could reverse the direction of an electric current using a device that he called the commutator, in 1833 he and Weber strung a double wire 8,000 feet long across the rooftops of Göttingen running from the observatory where Gauss lived and worked to the physics lab where Weber worked. Gauss and Weber developed an alphabet and were able to send messages, first individual words and then complete sentences. The first message, *"Michelmann kommt,"* indicated that the messenger Michelmann was on his way to the physics lab.

Their code consisted of up to four pulses that were either to the right or to the left. In that sense it is similar to the Morse code (which consists of a series of long and short beeps or flashes of light), which was devised about ten years after the Gauss/Weber code, and the binary code, which is used in modern computer programming. In all these cases, for each unit there are only two possibilities, left or right, short or long, and 1 or 0.

The telegraph line lasted 12 years until it was destroyed by lightning in 1845. The lightning was evenly distributed along the entire line, and as a result the pieces (some fairly long and some as short as four or five inches) showered on the town below it. The only recorded damage to anything were a few small holes in a lady's hat.

Gauss and Weber knew that their telegraph had great potential in a world where communication was becoming more and more important. They suggested using the rails of the newly invented railroads instead of one or both of the wires, but that suggestion was dismissed as too expensive. The Morse code was first used along the railway line between Baltimore and Washington, D.C., just as Gauss and Weber had recommended. Because information about the Gauss/Weber system was printed only in German, it escaped the notice of people outside of Germany. In fact, Gauss and Weber have been mostly forgotten for their role in the development of a system of telegraphy. If they had set it up in a large city instead of a small university town and if they had fought for recognition of their system, the history books might remember it. They were scientists, not entrepreneurs, and once they had succeeded they were ready to move on to the next challenge.

The Göttingen Seven

In 1837 the University of Göttingen, at that time the premier university in Europe, celebrated its 100th anniversary. Humboldt had to decide whether to accept Gauss' hospitality and attend the hundredth anniversary of the founding of the university at Göttingen, or to attend a convention of scientists in Prague. He wrote to Gauss, "I have to choose between several hours with you and several days of dull lectures by scientists who are most interested in impressing one another and eating well. I accept your invitation with the greatest pleasure."

Humboldt came to the celebration in Göttingen and the two scientists enjoyed one another's company greatly. Scholars from all over Europe attended the festivities. Gauss wrote at this time that "a flood of poems" broke out! Although the university bestowed honorary degrees in a variety of fields, Gauss was disappointed that his friend and colleague, the mathematician Sophie Germain, did not live long enough to accept hers. She had died of cancer six years earlier.

Gauss, recognized as the "Prince of Mathematics," was chosen to give a lecture to the Royal Society of Sciences as part of the festivities. He spoke humbly, offering his speech as a modest gift to the university where he had flourished for so many years. He was aware, however, that not everyone wanted to hear a rarefied scientific lecture. "I understand that most people do not dream of hearing a lecture on the abstract sciences. In addition to that, I do not pretend to be an impressive speaker. However, I would suggest that even if my topic seems dry to you, you will understand nothing if you do not listen. If, on the other hand, you give me your attention, you might find it interesting."

The audience was enthusiastic, and Gauss was indeed the hero.

Through all the activities of his life, Gauss continued to tend to his mother, who by this time was blind and feeble, but nonetheless mentally sharp. One afternoon in 1839, he and his 97-year-old mother sat in their living room in the observatory, sipping tea.

"*Mutter*, can I pour you some more tea?" Gauss asked.

"Yes, Carl, that would be lovely. You know, son, I have lived a long life. When you were young, I sometimes wondered if I could possibly keep up with you."

"Yes, *Mutter*, you have stayed with me through it all."

"Carl, do you remember the time the Duke's carriage splashed us as we were walking to the market? You were very little—not more than four years old, as I remember."

"Yes, I remember, *Mutter*."

"Do you remember what you said that day?" she asked.

"Not really."

"Well, you were talking about your future. You talked about some things that you wanted in life. The first thing that you wanted to do was to meet the Duke. You did that when you were 11 years old. Next you wanted to be a prince, like the Duke."

"Yes, and you told me there was no way I would ever be a duke because my father was not a nobleman," said Gauss.

"Yes, but now I hear people call you the 'Prince of Mathematics!' I think that means that you are a prince," she said proudly.

"Yes, people call me that. I'm not sure I deserve it, though," Gauss observed.

"The next thing you said that day was that you wanted to be rich," she continued. "I can't believe how comfortable we are in your home here, Carl. You have been successful in all that you have done."

"If you mean being called a prince and having enough money is the sum total of my accomplishments, I have to disagree with you, *Mutter*," said Carl. "The big thing I wanted to do was to study mathematics so deeply and continuously that I could have the pleasure of making many exciting discoveries. I don't care if they call me 'Prince.' I don't care if I have enough money. What I care about is that I have had a successful career as a mathematician. The Duke of Braunschweig is responsible for that. Without him, I would never even have made the first step."

"Yes, Carl, but you have been fortunate in other ways too. Your first wife Johanna was a perfect woman."

"Yes, she was. I didn't know how to continue after she died, but you know Minna was a fine wife too," Carl explained. "And the children were a pleasure to watch grow up. I still regret little

Louis, but there is nothing to be done for him now. I wonder if doctors will learn enough that they can help mothers like Johanna and babies like Louis in the future."

"Carl, we can't expect life to be sunny and bright all the time," said his mother.

"I know that, *Mutter*. Isn't it ironic, though, that after all that has happened, it is you and I who sit here looking back over our lives. My life started with the two of us together, and we are still together. I hope you look back with some pleasure on your own life."

"Yes, Carl, I do," she said, "but much of that is because of your success. I dedicated myself to you, my only child, the day you were born, and you have not disappointed me."

When Gauss' mother died in 1839 at the age of 97, Gauss described his mother's life as full of thorns, but he did what he could to make her last years as happy and comfortable as possible. He certainly had not died young and abandoned his mother as the gossiping neighbors had predicted when he was young. He outlived her by 16 years.

Gauss' older daughter Minna was active in the social life of Göttingen, particularly after her stepmother's death, and in 1830 she married Georg Heinrich August von Ewald, a professor of theology and oriental languages at Göttingen. Gauss, delighted with his son-in-law, enjoyed both his professional and personal relations with the Ewald couple as long as they remained in Göttingen. Unfortunately, it was only a few years after their marriage that his daughter Minna began to show symptoms of tuberculosis, quite probably the result of nursing her stepmother for so many years. In many ways she seemed to be the brightest and most capable of Gauss' children, and she was clearly her father's favorite.

In December 1837 soon after the centennial celebrations, another problem arose in the Ewald and Gauss households: the "Göttingen Seven" were fired. These seven professors, among them Minna's husband, Ewald, Gauss' physics professor and collaborator, Weber, and the brothers Grimm of fairy tale fame, had signed a petition demanding that the new king of Hannover revoke his proclamation replacing the constitution of 1833 with the earlier constitution of 1819. The professors, arguing that the constitution of 1833 was the official constitution and that the king couldn't choose to ignore it just because it didn't suit him, said that they had sworn their allegiance to the constitution of 1833. The king's action could not release them from this oath.

They knew the king's reason for the change: the 1833 constitution forbade an heir to the throne to become king if he had a physical or moral defect. The new king's only son was blind, although he had tried to keep that information private. The king was particularly angry that the petition was not sent privately to him—instead it was published throughout the community and state and beyond. The university officials tried to get the professors reinstated or to convince the professors to make a statement indicating that they hadn't really meant it, but the seven professors were adamant. Three of the professors were banished from the state of Hannover within three days, while the others could remain in Göttingen without pay as long as they "behaved" themselves.

The current Duke of Braunschweig was incensed and threatened to reopen the university at Helmstedt and hire the professors there. When nothing came of that, all the professors found positions in other universities. Gauss' daughter Minna moved with her husband to the university at Tübingen. Gauss tried vainly to convince the king to allow Weber to remain in Göttingen so that they could continue to collaborate on magnetism, but his request was denied. Gauss did not try to intervene on behalf of his son-in-law Ewald because he did not approve of nepotism. Ewald taught for ten years at Tübingen, and Minna died there, causing her father grave sorrow.

After the revolution of 1848, the government of Hannover tried to undo the unpleasantness of 1837 and it invited the seven professors to return to their places in Göttingen. Only Ewald and Weber returned. Although Gauss rejoiced in their return to Göttingen and enjoyed their friendships, he was aging and was never able to work with Weber the way he had earlier.

In 1840 Gauss' oldest son Josef (then 34 years old) married the daughter of a physician in the nearby town of Stade. In 1849 they had a son, Carl August Adolph Gauss. Once when the young family was visiting in Göttingen, Gauss asked his young grandson, "Well, my boy, what do you expect to make of yourself?"

The child replied, "Well, what do you expect to make of yourself?"

Gauss' answer was perfect: "My boy, I am already somebody."

Josef started out in the army, but eventually he left the army for a career with the railroads. He remained in touch with his father, but visited him only occasionally. He certainly lived up to Gauss' wish that he might have a career in the new technologies. A very practical and successful man, he soon was playing a major role in the railroad in Hannover. The railroads—carriages on land that were propelled with steam power—had always fascinated Gauss, and in 1854 he made his last trip away from Göttingen to view the construction of the railroad line between Göttingen and Kassel, through countryside that he and Josef had covered by coach many years earlier on their trip to Munich.

Therese had been acting as her father's housekeeper for several years, and they occasionally invited professors from the university to eat with them. Although Gauss was not the caricatured absentminded professor, he was sometimes amused by his eccentric colleagues. One evening, law professor Georg Julius Ribbentrop joined them for supper. A torrential rain began during the meal, and, when it looked as if it was going to continue to rain all night, Therese suggested that *Herr* Ribbentrop spend the night at the observatory. He would surely get soaked if he were to walk home through the downpour. The professor accepted the invitation graciously, but soon disappeared, only to return half an hour later, dripping wet, explaining that he had gone home to get his nightshirt!

Another time Gauss invited the same Professor Ribbentrop to visit at the observatory to view an eclipse of the moon. Since once again it turned out to be a rainy night, Gauss assumed that the appointment was off. When Ribbentrop appeared at the appointed time anyway, Gauss explained that, alas, it would not be possible to see the eclipse because of the rain. Ribbentrop replied, "Not at all, my dear Gauss! My landlady has seen to it that this time I did not forget my umbrella!"

Through her mother's final illness, her grandmother's blindness and eventual death, and her father's gradual slowdown of activity, Therese served as his hostess and companion. Gauss was utterly devoted to her, and the feeling was reciprocated. Therese did not marry until after her father's death. Neither she nor her sister Minna had any children.

Looking Back

During the last period of his life from 1841 to 1855, Gauss spent much of his time on pure mathematics, particularly analysis and functions of complex variables. His last scholarly work was his fourth proof of the fundamental theorem of algebra, which he had first proven in his doctoral dissertation in 1799. He also pursued his astronomical observations and was one of the first people to observe the planet Neptune soon after its initial discovery in 1846 by Galle, one of Gauss' friends who had told him of his discovery before publishing it.

In 1849 friends and colleagues at Göttingen decided to celebrate the 50th anniversary of Gauss' doctoral degree. When Gauss heard about the plan, he was outraged. He had simply used his genius to explore the world's most pressing problems with his mathematical skills and understanding. Isn't that what a scholar is expected to do? Undeterred, the authorities quietly proceeded anyway, and the resulting celebration was a wonderful reunion in Gauss' honor with many of his colleagues and friends in attendance. He was dumbfounded. He was granted many awards, including honorary doctoral degrees and honorary citizenship in

Braunschweig and Göttingen. The king himself wrote a congratulatory proclamation in his own handwriting. The high point of the celebration was when Gauss delivered a scientific lecture to a standing-room-only crowd. Gauss was disappointed that none of his sons were able to attend the festivities, although his daughter Therese acted as his hostess once again. Josef couldn't attend because of his responsibilities with the railroad at Hannover, and Eugen and Wilhelm were unable to make the long trip from America. Everyone agreed that Gauss was the world's greatest living scholar.

In his later years, Gauss was called upon to help with the widows' fund of the university. This fund, which was supposed to care for the widows and orphans of professors, was close to bankruptcy when Gauss took over. He analyzed the situation, used the laws of probability for his calculations to predict the expected life spans of the people involved, and was able to reorganize the fund in such a way that it could always take care of the aged women and orphans in its charge. In this project Gauss laid the foundations for the mathematics of modern insurance practice.

For years Gauss had played cards regularly with a group of friends. The games were certainly just for fun, but Gauss kept meticulous records over the years of how many aces each person had in each game. It gave him satisfaction to see the rules of probability play out week after week.

As an adult, he had a notebook of lists, including the number of steps from the observatory to the physics laboratory, the number of days his friends and famous people lived, and the ages of his children when they took their first steps or cut their first

Stamp celebrating Gauss' 200th birthday in 1977

tooth. He wrote a special letter to his friend Humboldt on the occasion of Humboldt's having lived precisely as many days as Newton. Three days before Gauss died, he calculated the number of days he had lived up until that day. Numbers of all kinds still fascinated him.

When Gauss died in 1855 at the age of 77, he was a very wealthy man. His basic salary of 2,500 *Thalers*, plus any fees he received from his students and the 110 *Thalers* he was paid for his role as assistant director of the Royal Society of Sciences, should not have yielded an estate of 152,892 *Thalers* in bonds plus the 17,965 *Thalers* that were found stashed in various places in his home. Yet there is no question of his honesty. He was a frugal man and a wise investor. He used the same genius in conducting his private finances that he applied to his mathematics.

Gauss' friend Sartorius von Waltershausen described Gauss' physical needs: "From childhood to maturity to old age, he was

10-Deutschmark note, front and back

always the genuine, straightforward Gauss. A small study, a table with a green cloth, a white standing desk, a narrow sofa and, after his seventieth year, an armchair, a shaded lamp, an unheated bedroom, plain food, a dressing gown, and a velvet cap—he never wanted anything more."

Simple though his needs were, Carl Friedrich Gauss was a great man and a great mathematician. His highly convoluted brain, preserved in formaldehyde, is one of the treasured relics in the medical clinic at the University at Göttingen. Gauss' brain ranks as one of the greatest repositories of human genius. He lived a total of 28,422 days of intensive mathematics. He was indeed the "Prince of Mathematics," or *Princeps Mathematicorum* in Latin, as he would have preferred to be called.

Index

17-gon, 86–88, 93–94, 98–99, 105, 146
50th anniversary of Gauss' doctoral degree, 233

algebra, 35, 42, 53, 57, 64, 65, 104, 117, 165, 233
Amiel, Henri, 47
Archimedes, 67, 93, 96, 118, 141
Ascension, 107, 111
astronomy, xii, 125–130, 138, 146, 153, 157, 163, 165, 187, 233
Auerstedt, battle at, 140

Büttner, J. G., 25–28, 33–38, 52, 57, 66, 67, 153, 215
Bartels, Martin, 27, 35, 37, 41–47, 51–57, 64, 121, 216
Benze, Friedrich, 4, 6–8, 13–18, 21, 28, 137
Berlin, 127, 201, 217

birthday, Gauss', 107–111, 140
Bode, Johann, 151
Bolyai, Johann, 213–214
Bolyai, Wolfgang, 82–93, 101–112, 120, 123–124, 133, 135, 213–214
brain, 119, 236–237
Braunschweig, 14, 51, 56, 67, 81, 83, 96, 98, 105–106, 139, 151, 160, 179
  Duke of, *see* Duke of Braunschweig
Brunswick, *see* Braunschweig

calculate, 33–35, 66, 101–103, 107, 127, 196, 216, 221, 235
canal, 11–13, 16, 17
card games, 152, 234
carriage, 21, 29, 56, 140, 166–170, 188, 201, 226
  iron-clad wheels, 169
  isinglass windows, 166

castle, 168
cathedral analogy, 98
Ceres, 125, 126, 129, 138
clock, 130, 131, 154, 172
Collegium Carolinum, 51, 54, 75, 78, 83, 87–88
commutator, 222
copper, 219

d'Alembert, J. LeRond, 43, 121
damask, 13–18, 21, 29, 137
daughter
 first, *see* Gauss, Wilhelmine (Minna)
 second, *see* Gauss, Therese
decimal expansion, 101–102
Dedekind, Richard, 145
degaussing, 221
Descartes, René, 79
dialect, 54, 66–67, 151
diary, *see* journal
difference of two squares, 34, 55
Diophantus, 79
Discourses on Number Theory, xi, 122, 141
distance formula, 65
doctoral degree, *see* PhD
Dransfeld, 192, 196
drawing instruments, 66, 86, 93, 98
du Châtelet, Emilie, 84–85

Du form of address, 104
Duke of Braunschweig, xi, 14–18, 21–23, 28–30, 38, 51, 56–61, 63, 66, 77–79, 88, 117–119, 122, 123, 128–129, 135, 139–141, 143, 226, 227, 229

Easter, 107–111
electricity, 222
ellipse, 125–126, 136
engagement
 to Johanna Osthoff, 135
 to Minna Waldeck, 159
Eschenburg, Wilhelm, 87
Euclid, 41, 58, 66, 76, 79, 82, 87, 93, 215
Euler, Leonhard, 42, 43, 67, 79, 87, 91–92, 104, 118, 121, 127, 140
Ewald, Georg Heinrich August von, xiii, 228–230
Ewald, Wilhelmine (Minna) von, *see* Gauss, Wilhelmine (Minna)

father, *see* Gauss, Gebhard
Ferdinand, Karl Wilhelm, *see* Duke of Braunschweig
Fermat primes, 86–87, 91
fiftieth anniversary of Gauss' doctoral degree, 233

INDEX

Finnish, 215
Fuggerei, 171
fundamental theorem of algebra, 42, 65, 104, 117, 120–121, 165, 233
fundamental theorem of arithmetic, 117, 122

Göttingen Seven, 229
gauss (unit of magnetism), 219, 221
Gauss, Carl August Adolph (grandson), 230
Gauss, Dorothea, xi, 4, 19–21, 31, 38–40, 60, 63, 64, 105–107, 115, 129, 149, 151, 159, 179–183, 191, 194, 199, 206, 207, 226–228
  illiteracy, 6, 20, 54, 160, 181
Gauss, Eugen, 161–163, 173, 200–201, 203–208, 234
Gauss, Gebhard, xi, 3, 13, 18, 22, 31, 36–37, 53, 57, 64–66, 108, 119–120, 123, 150, 160, 204
Gauss, Johanna, *see* Osthoff, Johanna
Gauss, Josef, 137–138, 145, 147, 149–150, 152, 161–163, 166–171, 175, 181, 190, 198, 206, 230–231, 234

Gauss, Louis, 154–157, 228
Gauss, Minna (wife), *see* Waldeck, Minna
Gauss, Therese, 172, 173, 180–183, 199, 201, 206, 207, 231–232, 234
Gauss, Wilhelm, 161–163, 201, 208–209, 220, 234
Gauss, Wilhelmine (Minna), 149–150, 161, 162, 206, 207, 228–230, 232
genius, 36, 37, 46, 47, 53, 58, 83, 116, 117, 119, 122, 129, 130, 134, 218, 233, 235, 237
geodesy, xii, 187, 197
geometry, 38, 41, 53, 57, 64, 76, 83, 97, 197, 213
  non-Euclidean, *see* non-Euclidean geometry
Georg IV, 165, 190
Germain, Sophie, 141–142, 154, 225
godfather, *see* Ritter, Georg Karl
Gould, Benjamin Apthorp, xii
greatest common factor, 102, 103
Greek, 41, 43, 64, 77, 89, 93, 127, 151
Grimm, Jacob Ludwig Carl and Wilhelm Carl, 229
Gymnasium, 51, 56, 57, 60–61, 63–71, 78, 88

Harding, Carl Ludwig, 154
hawk, 168–170
heliotrope, 190–191
Helmstedt, 78, 89, 96, 116, 118, 120, 131, 143, 229
Herr Hofrat, 165
High German, 66–67
Hils, 192
Hohehagen, 192
housing project for the poor, 171
Humboldt, Alexander von, 217, 219, 225, 235

Ide, Johann, 64–65, 71–73, 87, 90–93
infinite series, 43–47, 165
infinite sums, 44
Institut de France, 154
insurance, 234
irrational, 58–60
irrational numbers, 61

journal, 95–98, 213

Kästner, Abraham Gotthelf, 84–85, 88, 90, 105, 120, 151
Knight's Cross of the Guelph Order, 165

Lüneburg Heath, 191, 197
Lagrange, Joseph Louis de, 67, 121
Lalande Medal, 154
Lambert, Johann, 68
lamp, 32, 38–40, 236
Laplace, Pierre Simon, 142
Latin, 38, 55, 64, 65, 77, 96, 97, 116, 154, 237
Latin of the modern world (English), 154
latitude, 76, 131
least common multiple, 102–103
least squares, method of, 72–73, 97, 125, 196
LeBlanc, *Monsieur*, *see* Germain, Sophie
Lecoq, Colonel K. L. E. von, 187
Legendre, Adrien-Marie, 73
Leibniz, Gottfried Wilhelm, 43, 84, 140
Lessing, Gotthold Ephraim, 67
lists, 234
Lobachevsky, Nicolai, 214–216
logarithms, 66, 69
longitude, 76, 131

Möbius, August Ferdinand, 145
magnetic association, 221
magnetic laboratory, 219–221
magnetism, xii, 219–221, 230
map, 131, 187–188, 195, 214, *see also* surveying
Marcellus, 141

marriage
  to Johanna Osthoff, 134–137
  to Minna Waldeck, 159–163
Mayer, (Johann) Tobias, xii, 217
measles, 200, 201
measurement, 71–72, 131, 132, 187, 190, 195, 201, 214, 219, 220
memory, 46, 92, 208
Mersenne primes, 86
method of least squares, *see* least squares, method of
modular arithmetic, 112
Monsieur LeBlanc, *see* Germain, Sophie
Monthly Correspondence of Mapping the Earth and the Sky, 125, 129
Morse code, 222–223
mother, *see* Gauss, Dorothea
motto, 97
Munich, 166, 171–172, 175, 231
myopia, 20, 130

Napoleon, 139–143, 191, 196
near-sightedness, *see* myopia
nepotism, 230
Newton, Isaac, 43, 68–69, 76, 79, 84, 98, 127, 147, 235
non-Euclidean geometry, 75, 97, 213–215

North Pole, 221
Notizenjournal, *see* journal
number theory, 65, 96, 101–112, 117, 121–123, 141, 165

Olbers, Wilhelm, 126–131, 142–143, 149, 156–157, 202, 208, 219, 222
  poem, 129
Onkel Friedrich, *see* Benze, Friedrich
optics, 165
orbit, 125, 136, 153, 154
origination point, 191
Osthoff, Johanna, 133–138, 144–147, 152, 154–157, 227

Pallas (planet), 127
Papen's Atlas, 188, 196
parallel postulate, 41–42, 75–76, 82–83, 213, 215, *see also* non-Euclidean geometry
Pauca sed matura, 97
period of the repetend, 101, 102
Pernety, General, 142
Pfaff, Johann Friedrich, 89, 118, 120, 121, 143, 146
PhD, xi, 118, 120, 123, 143, 195
  fiftieth anniversary of, 233
physical description of Gauss, 115
Piazzi, Giuseppi, 125, 137
pipe, 124, 151

postulates, Euclid's, *see* parallel postulate
Prime Number Theorem, 70
prime numbers, 54, 68–70, 91–92
Prince of Mathematics, 22, 30, 153, 225, 227, 237
Princeps Mathematicorum, 237
Principia, The, 68, 84, *see also* Newton, Isaac
probability, 234
prodigy, 3–9, 53–58, 149
Pythagorean Theorem, 65

Queen of Mathematics, *see* number theory
quill, 27, 30, 153, 181

rational numbers, 101
reciprocals
  of the natural numbers, 46
  of the positive powers of two, 44–46
  of the whole numbers, 101–102
regular 17-gon, *see* 17-gon
Reichenbach, Georg Friedrich von, 166, 172
remainders after division, 108–112
repeating decimal, 101
repetend, 101

Ribbentrop, Georg Julius, 231–232
Ritter, Georg Karl, 31–32, 81, 133
Ross, Captain James Clark, 221
Royal Observatory, 221
Rumpf, 189–190
Russia, 35, 126, 127, 139, 214
Russian (language), 84, 216

Sanskrit, 215
schedule, 122, 146, 165
school, 13, 25–28, 30, 33–36, 51, 63–71, 190, 203, *see also* Collegium Carolinum, *Gymnasium*
Schumacher, Heinrich Christian, xii, 187–190, 207, 214–216
seal, 18, 97
sextant, 130, 187, *see also* surveying
Seyffer, Professor, 81
son
  first, *see* Gauss, Josef
  fourth, *see* Gauss, Wilhelm
  second, *see* Gauss, Louis
  third, *see* Gauss, Eugen
South Pole, 221
spa, 131, 170, 175–176
spinning flax, 31, 37, 63
square root of two, 58–60

INDEX

squares, method of least, *see* least squares, method of
St. Petersburg, 126–128, 139
St. Petersburg Academy of Sciences, 126
straight-edge, 66, 86, 87, 93, 98
sum, 19, 33, 35, 46, 53, 73, 97, 208, 215
surveying, 187–193, 206, *see also* measurement

tablecloths, *see* damask
teaching, 143, 144
telegraphy, xii, 222–223
telescope, 130, 139, 172, 189
terminating decimal, 101
tesla (unit of magnetism), 219
theodolite, 201
Theory of the Orbits of Celestial Bodies Moving in Conic Sections, The, 153–154
thesis, 118, 120, 121
Tittel, Paul, 166, 175
top, 32
triangular number, 96–97
triangulation, 131, 190, 195, *see also* surveying
tuberculosis, 175, 200, 205, 207, 228
turnip, 38–40

Uncle Friedrich, *see* Benze, Friedrich
Utzschneider, Joseph von, 166

velvet cap, 149, 193, 197, 236

Würzburg, 171
Waldeck, Minna, 152, 159–163, 166, 171, 173, 198–201, 204–207, 218, 227
Waltershausen, Baron Wolfgang Sartorius von, 153, 235
water power, 150
weaving, *see* damask
Weber, Wilhelm, xii, xiii, 217–223, 229, 230
Weende, 192
wet-nurse, 155
widows' fund, 234
Wilkes, Charles, 221
World War II, 221
writing, 145, 153

Zach, Franz Xaver von, 125, 130, 139, 187
Zeno's paradox, 43–46
Zimmermann, Eberhard August Wilhelm, 51–58, 66, 67, 75–78, 84, 88, 98–99, 128, 144